CABLE
TELEVISION
HANDBOOK

Cable
Television
Handbook

Eugene R. Bartlett

McGraw-Hill
New York • San Francisco • Washington, D.C.
Auckland • Bogotá • Caracas • Lisbon • London
Madrid • Mexico City • Milan • Montreal • New Delhi
San Juan • Singapore • Sydney • Tokyo • Toronto

Library of Congress Cataloging-in-Publication Data

Bartlett, Eugene R.
  Cable Television Handbook / Eugene R. Bartlett
    p. cm.
  Includes bibliographical references.
  ISN 0-07-006891-7 (book)
    1. Cable television—Handbook, manuals, etc. 2. High definition television—Handbooks, manuals, etc. 3. Color television—Handbooks, manuals, etc. I. Title.
  TK6675.B36 1999
  621.388'57—dc21   99-043784

McGraw-Hill

A Division of The McGraw·Hill Companies

1 2 3 4 5 6 7 8 9 0   DOC/DOC   9 0 9 8 7 6 5 4 3 2 1 0 9

ISBN 0-07-006891-7

This book was set in Century Schoolbook by D&G Limited LLC.

Printed and bound by R. R. Donnelley & Sons Company.

McGraw-Hill books are available at special quantity discounts to use as premiums and sales promotions, or for use in corporate training programs. For more information, please write to Director of Special Sales, McGraw-Hill, 11 West 19th Street, New York, NY 10011. Or contact your local bookstore.

This book is printed on recycled, acid-free paper containing a minimum of 50 percent recycled de-inked fiber.

CONTENTS

Contents

About the Author

Eugene R. Bartlett is a cable industry consultant, Instructor at the ITT Technical Institute, and independent technical writer. He was formerly the director of engineering and partner in several cable television systems. Bartlett is the author of the first edition of this book, *Cable Television Technology and Operations*, and *Cable Communications*, both published by McGraw-Hill.

Preface

The material covered in this book is essentially an update of the growth and advancement of the cable television industry since Cable Television Technology and Operations was published in 1990. Much has transpired within the industry in both a technical and a regulatory sense in the past decade. In many instances, cable operators have manipulated the position of their networks to take a place in the wired telecommunications infrastructure, while many operators have done little or nothing. For some cable television systems, this has been easier than for others, depending on the network architecture and quality of system components. Essentially, the job at hand is to ready the return path for active two-way operations, as required for telecommunications applications. This is not easy to achieve, and for many systems, it is extremely difficult and costly. However, for a cable television system to become a cable telecommunication system, it must be accomplished.

Today many of the systems owned by some of the major MSOs are operating as cable telecommunications systems and they are successful mainly because the active return path was designed, built, and activated at the time of the initial build. Most of the readers of this work, I am sure, are well aware of the various problems with noise and signal ingress in the return path, as well as some of the solutions. I recall my first spectrum analyzer presentation of my first active subsplit reverse system. There was WWV at 5 and 10 MHz sticking up like poles, plus a myriad of other carriers and a significant level of noise on the screen. Since there was no immediate application for use of the return path, it gave me and the technical staff time to work things out. It was a difficult, meticulous job to tighten and balance the system. Choosing the quiet and vacant spots in the 5-to-30-MHz reverse band enabled the use of a few carriers for the return path applications needed at that time.

Cable television systems are based on cascaded amplifier theory and this book will review this subject in a practical, detailed manner. Mathematical proofs and tutorial subjects appear in the appendices, as noted in the text. In general, detailed tutorial subject matter will appear in the appendix so the general flow of the work is not interrupted. Readers who feel the need for tutorial work will no doubt find the material in the appendices helpful.

Network topology and design are discussed in detail with the application of fiber-optic technology. Thus, the number of amplifiers in cascade are decreased, improving the signal-to-noise ratio and signal-to-distortion ratios. For the systems using these techniques, the signal and picture quality have been improved. In many instances, cable television systems have improved the picture quality and the number of channels for the forward downstream system. Still, for a great many systems, the return path needs similar treatment so that two-way telecommunications services can be offered to subscribers. Many systems lack any meaningful business data so that cash flow and proforma studies can be made. Until business projections for services requiring two-way communication are made, many cable systems are adamant about investing time and money in restructuring and readying the return path for telecommunication applications. In several instances, some major system operators have gone this route and are pointing the way as to the problems and benefits.

This book summarizes the basic cable television growth to the present day, analyzes the design and implementation problems relating to plant upgrades, explores fiber-optic and digital techniques, and addresses the subscriber terminals and system interfacing problems. Plant maintenance and testing is covered in the last chapter where state of the art instrumentation and methods are used to maintain a modern cable telecommunication system.

Acknowledgments

Again, I want to thank my wife, Mona, for her help and encouragement in preparing the manuscript. I also want to thank my son-in-law, Daniel Begeman, for his work in preparing the graphics and working with a tight time schedule. Also I wish to thank my son, Russell, for his help and advice in preparing the manuscript.

A special thanks to Mr. Gene Faulkner of Hukk Engineering for permission to include information on the new QAM monitor instrument. Also again, a special thanks to Mr. Peter Harper of Hewlett-Packard (Cerjac) for including a study of SONET testing methods in Appendix C.

Introduction

Cable Television History

The reason cable television systems were developed in the first place was to give improved television service to people living in remote areas where the off-air signal was poor. The television broadcasting stations looked on this new industry as an ally in expanding their market area. It should go without saying, even today, that the local broadcast stations are a major source of local news, weather, and sports to the American viewing public. Indeed, cable television still carries the local television broadcast stations as part of their channel line-up.

It should also be evident, for the most part, that the reception problem for many TV broadcast stations has not disappeared. Transmitted power output, as stated in the station licenses, has not changed. The location of the station's transmitting antenna towers, as well as the antennas themselves in most cases, has not changed. The point is that the off-air reception problem remains. Therefore, carriage of local TV broadcasting stations by cable television system operators is still a good reason for many subscribers to be connected to the cable television system. Satellite operators such as C-Band/Ku Band and the recent DBS Satellite Systems provide no local programs, however. Since many early cable television systems had to provide locally originated programming written as a requirement in their license, charter, or franchise, they provided another service to the subscriber for local news, weather, and sports. Many a family were able to watch their children participate in school sport programs televised by the local cable company, yet only cable subscribers could see these programs. Many cable operators looked upon their locally originated cable channel as a horrible problem and large expense. Some cable operators became successful enough to sell local advertising spots to offset much of the costs of producing the programs.

Today the cable television industry is facing much competition for the viewing public's consideration. Analysis of the facts tells us that a family that has a DBS or some other satellite service has to have some kind of roof-top antenna or the usual rabbit-ear antenna to provide the local television off-air broadcasts. Many new housing developments and planned communities prohibit backyard dishes and or roof-top antennas. This is all the more reason for the viewing public to remain on a cable television system. We, the cable operators, must never forget this fact.

The Television Reception Problem

The reception problem of the off-air television broadcast stations remains the same as it did when the cable television industry started. As stated earlier, the problem has not simply disappeared with technical progress. Of course, the television set itself has been improved significantly. The development of solid state components and the manufacture of printed circuits allows sets to run cooler, thus prolonging the set life. The tuner circuits have improved signal-to-noise ratios and the mechanical switch has been replaced by push-button digital switches. Remote controls have helped to make so-called couch potatoes of many people. Large screens and improved sound have made modern TV sets an important source of home entertainment.

Remote Areas—Separated By Distance   Remote areas separated by large distances are places where cable television is the major source of television programming. Such areas usually have a sparse population, so subscriber numbers are usually small. Cable systems serving such areas provide a very useful service and have to charge subscribers accordingly. Since the path loss of the signal between the TV transmitter and receiver is a function of distance and frequency, remote areas separated by large distances receive weak TV signals.

Effects of Terrain On Reception   The effects of terrain on reception also affect the signal levels. Mountainous areas cause problems with television reception in many parts of the United States. It has been reported in many books and magazines that the first cable television systems were started in mountainous territory. People living in the valleys, where most of the towns were built, also had major problems with television reception. Essentially, these locations were out of the field-strength path as projected by the transmitting sites. In many instances, homes that were fairly close to the transmitter were nearly completely shielded by a mountain in-between. Cable systems serving the areas enjoyed a high penetration of television service.

Effects On Reception By Dense Buildings In Cities   Many residents of cities also have television reception problems. Tall buildings and skyscrapers act similarly as mountains and shield many apartment dwellers from good quality reception. In many instances, television signals are reflected off some of the surrounding buildings, causing severe television picture impairment, which is called multipath reception. Multipath reception can be so severe that a home can find that the signal for a given television station is unwatchable, while for another channel, it is satisfactory. Such multipath problems are illustrated in Figure 1-1.

Early Cable Systems

Now that the need for cable television systems has been addressed, a brief study of the development of the industry will follow. The technique used by early system operators was to start with good television off-air signals, usually received from a mountain top or a tower or water tank location, and connect subscribers to this source by a cascade of amplifiers connected together by sections of coaxial cable. Surplus coaxial cable from WWII and home-built RF amplifiers made up the basic components of these early systems.

A Community Antenna System   The foregoing describes the system for what it actually was, a *Community Antenna Television* (CATV) system, referred to in present day vernacular as classic cable. Such so-called start-up systems followed no specified procedures in connecting the system together. Technicians and engineers were usually fresh out of the armed services. Engineers who graduated from engineering schools were taught

Figure 1-1
Multi-path reception

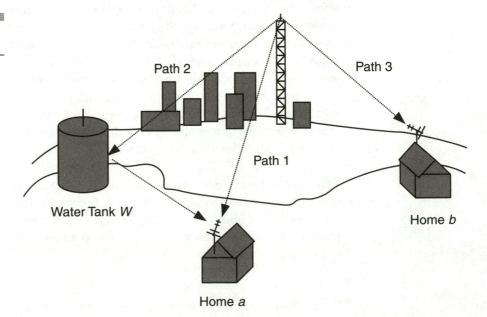

electronics as radio technology and engineering as power systems. Writings from the telephone company sources such as AT&T/Bell Labs produced early theory on cascaded amplifiers. With such little information to go on, it was the resourcefulness and determination of the early system technicians and engineers that caused the industry to grow to what it is today. For many of us, our first systems were essentially classic cable systems that provided vastly improved broadcast television to our subscribers.

Single Cable, Low Television Bands Most early classic systems consisted of simply a cascade of amplifiers connected together by sections of 75-ohm cable. Connectors were used for that type of cable with the matching connectors on the amplifier housing. Matching networks were used to connect subscribers via a drop cable. The matching network tapped a signal off the main cable feeding signal to the connecting drop cable. Amplifiers were usually either home-built or used whatever was available on the newly formed MATV industry. Power for these first amplifier types was often 110 volt a.c., thus requiring the power company to provide a power source. It goes without saying that technical personnel faced definite safety hazards working on these start-up systems.

The attenuation of the cable was not as good as cable used today. Since the loss was much less at the low VHF bands, the upper VHF channels and any UHF channels were converted to the low VHF band of channels two through six. This technique allowed more distance between amplifiers and thus provided more "reach" for the system. Many early systems were essentially single-cable, five-channel systems that enabled the start-up of the cable television industry.

Early System Improvements, Cable, Electronics The braided-type cable gave many system operators a lot of trouble, mostly with water ingress. Of course, early systems leaked signals, but it was not a large problem since the signal leakage was "on channel". Noise ingress and the accompanying impedance mismatch was much more of a problem with system signal levels as well as picture ghosting and noise. Clearly better cable was needed and since the supply of government surplus cable became scarce, cable manufacturers eager for new business made improved cable for the fledgling industry. The solid aluminum-sheathed, plastic foam-filled cable fulfilled the cable television industry's needs. Several sizes of cable with improved loss characteristics allowed cable systems to extend the aerial plant longer distances, serving more subscribers. Along with the cable, connector manufacturers were able to supply improved connectors, which exhibited improved electrical and weatherproof characteristics.

About the same time, several manufacturers started making RF cable amplifiers with vastly improved noise figures, distortion characteristics, and improved operating control features. It was these improvements that caused the industry to grow at a rapid rate. The pressure tap was introduced with the arrival of the solid aluminum-sheathed cable. This allowed a better method to connect subscribers to the system. These pressure taps exhibited an improved impedance match to the cable as well as better weatherproof characteristics.

The Early Systems Evolve

Indeed, early systems were growing both in plant size and the accompanying subscriber population. In many instances, these systems made the difference between watching television and not watching television. Since more and more community residents wanted to get connected to the cable system, cable operators had to extend the plant. Needless to say, the improved cable with its superior loss characteristics enabled the use of the VHF Hi Band, thus adding channels 7 through 13. The 12-channel system now gave cable operators more channels, thus causing an increase in subscriber interest to connect. System architecture consequently needed to be improved to cover a larger area serving many more subscribers. Once the industry became an obvious success, more technicians and engineers became involved, both at the system and the manufacturing level. It is interesting to note that in many instances, as early literature indicates, the manufacturers were telling the cable operators how to extend and improve their systems and what was needed to do so.

The shortage of trained technical people became more and more of a problem. Realizing this, the Society of Cable Television Engineers was formed in the late 1960s and early 1970s. The Society provided an exchange of technical information and practices among its members and eventually, through meetings and seminars, became the most important means of technical education. These technical meetings and seminars sponsored by the new society developed a great working relationship with many manufacturers and equipment vendors in providing both speakers and technical literature to the members. Even today, this society is important in the education of its members. The technical certification program developed by the now-called Society of Cable Telecommunication Engineers provides a realistic measurement of a member's technical expertise.

Separate Trunk-Feeder Systems One of the main developments of early systems was the evolvement in the network architecture of a cable television system. From the simple single cable system, the trunk feeder system was developed. The cascade amplifier theory indicated that if the signal level was preserved by limiting the losses attributed to subscriber taps, the overall distance or reach of a system was improved. At amplifier locations, a tap-off or bridging amplifier was added where a small amount of trunk signal was tapped off, amplified, and connected to a separate feeder cable system containing an improved subscriber tap. These taps were actually directional couplers with more than one output port. The main trunk cable carried a high-quality signal to the extremities of the cable system while the feeder cable carried the subscriber cable with the taps. Subscribers were connected to the system via a drop cable connected to one of the tap ports. This concept allowed the cable operating companies to build their plant farther from the signal source or head-end, passing more and more homes along the way. Higher and higher subscriber counts resulted in improved cash flow needed to support company operations and pay debts.

The Tree Network   This network topology resulted in what is called the trunk-feeder system or, more appropriately, the tree network. No subscribers were connected to the trunk cable, thus preserving the signal quality. Subscribers were only connected to the taps in the feeder network. Thus, the trunk and branch likeness to the tree resulted. The whole design process of such a system now is a lot different from the early single cable systems. This type of cable system is shown in Figure 1-2. Essentially, the trunk cable was a signal transportation system and the feeder network was the signal delivery system.

Cable Powering   Early system amplifiers were powered from the 110-volt a.c. line. Often a watt/hour meter was installed at one amplifier and its consumed power was monitored by the local power company. Often the other amplifiers were merely counted and the single amplifier consumption multiplied by the number of amplifiers was used to calculate the monthly power charges to the cable company. In some instances, the local power company required a meter at each amplifier location. Connections to the local power distribution system required the cable operators to properly ground and bond their systems according to the National Electric Safety codes. Of course, technicians working on the cable system also faced the possibility of electric shock.

The Dedicated Taps   Another improvement that had an immense impact on the cable television industry was the development of the directional tap. Also known as the subscriber tap, it provides a signal to the drop cable feeding the subscriber's television set. These taps were made with an input and

Figure 1-2
Example of trunk
feeder design

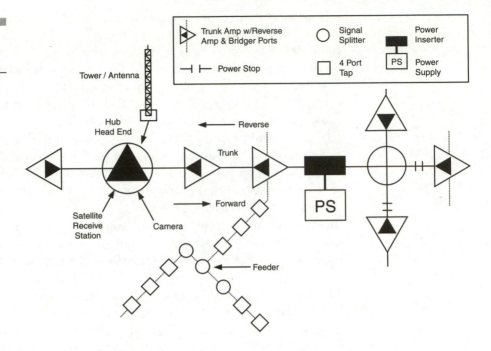

output coaxial connector for the feeder cable, plus two or more "F" connector ports for the subscriber connections. The "F" type connector and the "C" type connectors were spin-offs from the MATV industry. Today many people in the cable television industry wish the 75-ohm version of the BNC connector was adopted instead of the "F" type connector. However, present day systems use a vastly improved "F" connector which, when properly installed, works very well.

Changes in Plant Design and Development

The changes in plant design and network architecture were driven by the development of the devices and components from the various manufacturers and equipment vendors. Often it was this equipment that pointed the way to the possibilities of such network improvements. Across the board, cable, amplifiers, couplers, dividers, taps, connectors, and power supplies were vastly improved and they still are even today. Each of these devices will be examined in the rest of this chapter.

Development in Electronics

The development in electronics always seems to be a source of amazement. Early amplifying devices were the vacuum tube that consumed much more power than the present day solid state devices. The development of the transistor and its use as an amplifying device was welcomed in the cable amplifier arena. The transistorized amplifiers consumed less power, developed less heat, and provided lower noise and distortion figures. It was a definite improvement. Solid state technology progressed rapidly from the single discrete transistor to the integrated circuit consisting of the equivalent of several transistors as well as resistors and capacitors. Such so-called chip technology resulted in *integrated circuit* (IC) amplifiers for the cable television industry. These amplifiers provided more gain and a higher output level, as well as lower noise and distortion characteristics.

Cable Powered Amplifier   One large improvement in cable television technology was the use of cable-powered amplifiers. The cable itself was used to feed power to the electronic amplifiers. Coupling networks within the amplifier housings would separate the signal from the power, thus feeding each to their respective circuits. At first, the voltage level of 30 volts a.c. was connected to the cable via a transformer whose primary was connected to the 110-volt a.c. commercial power system. This 30-volt power would supply a segment of the cable system consisting of several amplifiers. Power blocks would disconnect the segment from other segments connected to other 30-volt supplies. These power blocks were bridged with signal-passing networks, thus providing signal continuity between power segments. It should be evident that now lower connections had to be made to the commercial power system. Only at 30-volt power supply locations were connections to the commercial power system needed.

This improvement now requires system designers to consider the power distribution problem along with the signal delivery system. Systems were always expanding and new plants were added, requiring more amplifiers and power supplies. It quickly became evident that increasing the cable power supply voltage to 60 volts would result in fewer line voltage drops and would allow more amplifiers per power segment. This voltage value falls within the National Electric Safety code as low voltage and thus was often considered as the optimum voltage value. Today, however, some systems use supplies with 90-volt a.c. as the cable supply voltage. In many of today's modern cable systems that employ fiber-optic technology, the fiber-optic system is powered from the coaxial cable section of the outside plant.

Solid State Transistorized Amplifiers   The solid state cable system amplifiers evolved from the transistorized version to the integrated circuit type. The main amplifier chip, as manufactured by several companies, was complete enough to be considered as an amplifier block. Companies using such chip technology merely had to supply the input/output and power connections, plus several control connections for amplifier gain and slope. In some amplifier integrated circuits, temperature compensation and circuit stabilization had to be provided. Proper heat sinking for these integrated circuits became part of the mounting method within the amplifier housing. This technology contributed to improved system performance as well as lower cost.

In later years, the bandwidth of such amplifiers provided proper gain and output level up to and including 1,000 MHz (1 GHz). This bandwidth of about 1 GHz allowed the operation of up to approximately 150 6-MHz NTSC television channels. It became quite clear that this large downstream bandwidth was indeed overkill, and other uses may, in the end, become more profitable. Thus, the cable telecommunication concept can make good use of the overall bandwidth.

The amplifier downstream delivery of television signals provided service basically "on channel," or on the frequency of television broadcast stations. No equipment was needed by the subscriber other than the normal television receiver. When programming was carried on the cable on non-standard television channels, a means of converting the programming to channels that a television set could receive was needed. The amplifiers had no problem with non-standard channels, but the television sets did. That, of course, is another story.

Surge Protection and Regulated Power Supplies   Since lightning and power surges caused power supply and amplifier failure, protective devices were developed. In older systems, fuses were the first line of defense followed by gas surge suppressors. Most systems in operation today use a variety of protective devices, but power system design can make a power supply more immune to transients and electrical surges.

First and foremost, the ferroresonant transformer with a winding supplying a large capacitive load provided a reasonably stable 60-volt a.c. supply. Usually the power inserter has protective fuses on the supply ports. At each amplifier location, a surge protector such as the gas variety or a *metal oxide varistor* (MOV) protected the amplifier power supply module from the cable system. Most modern cable amplifier power supplies were switching power supplies in which the voltage to the amplifier gain blocks (amplifier integrated circuit) was highly regulated.

The previously mentioned ferroresonant transformer power supply was first developed as a 30-volt a.c. power supply. As many systems grew larger, it became obvious that a higher voltage power supply was more applicable to the cable system powering needs. Hence, the 60-volt power supply became standard and remained so for many years. Now with many cable plants using fiber-optic systems powered from the coaxial cable system and coupled with an activated upstream system, a 90-volt supply obviously is needed.

Since plant reliability is extremely important in cable systems designed for two-way voice/video and data service, stand-by power supplies providing an appropriate amount of stand-by time are definitely required. Since the nature of ferroresonant transformer supplies requires the core to remain in saturation, the output becomes a quasi-square wave. The primary current has to keep the transformer core in saturation at light loads, which means the current drawn is significant. Thus, stand-by supplies use a different, less power-hungry method of regulation needed to provide a decent amount of stand-by time. It is, of course, desirable to have the a.c. power supply operating between 75 and 85 percent full load at least.

Automatic Gain Control   From experience with early systems, the need for automatic gain control was quickly recognized. Variations in amplifier output due to temperature changes occurred in both the cable and the amplifiers themselves. Thermal compensation of the amplifier gain helped greatly in the stabilization of the amplifier output levels, but the compensation of signal levels due to cable temperature changes required automatic gain control circuits. These circuits required two pilot frequencies located in both the low portion and high portion of the signal band. The pilot carriers were appropriately termed the high pilot and the low pilot. Level circuitry measuring the amplitude of both carriers responded by controlling the gain of the amplifiers at both the high and low end of the frequency band.

At one time, it was thought that alternating between controlled amplifiers and non-gain controlled amplifiers (manual gain controlled) was sufficient to maintain a constant output level. However, as greater bandwidths and higher levels were needed as systems expanded in distance and in the number of channels, every amplifier in a trunk cascade had *automatic gain control* (AGC), which then became the standard. As distance requirements for cable plants increased, better amplifiers were needed. Amplifiers that had high-level output, lower noise, and distortion specification were also developed. The feed forward amplifier improved third-order distortion specifications and was useful in long amplifier cascades where third-order distortion is a limiting factor. The power-doubling amplifier provided higher signal levels in long cascades and in feeder branches that contained a large amount of subscriber taps.

Many amplifier manufacturers provide a broad selection of amplifiers that aid system designers in designing efficient and economical cable plants. The power-doubling and feed forward-designed amplifiers enable coaxial cable trunk lines to be extended farther into the system. Cable system operators could install the reverse amplifier modules into sections of the coaxial trunk to activate the sections desired for two-way service. The necessary extra system power required to activate a trunk feed or return system should be designed into the system in the first place to avoid having to go out and upgrade the power system.

Slope Control, Equalization   The cable system amplifiers in present day systems are greatly improved over the previous type. Earlier amplifiers lacked proper thermal stability, causing the gain to shift. Thermal compensation devices that appeared in the newer designs took care of this problem. Since the cable characteristics also change with temperature, AGC became a necessary addition to the amplifiers.

The AGC method employed the use of a low-frequency and high-frequency pilot whose signal level was used as a reference level for the automatic control of a system's signal level. Coupled with the built-in thermal compensation, the overall temperature problem was under control. Now system signal levels for long trunk amplifier cascades were stabilized and subscriber signal levels did not vary.

As solid state technology advanced into the integrated circuit area, amplifiers in cable television systems improved immensely. Less power was consumed and thermal stability improved since much of the circuitry was contained in the same IC chip. Now with automatic gain and slope controls added to an improved amplifier, many cable systems could expand their bandwidth, thus increasing the number of channels offered to subscribers, which had a positive impact on cash flow. On the other hand, if it was more desirable to extend plant distance rather than expand bandwidth, the lower noise low-distortion amplifiers made this trade-off possible. Then with the arrival of optical fiber, the whole picture changed again and for the better. This subject is covered in detail in Chapter 3, "Fiber-Optic Technology in Cable Television Systems."

The Unity Gain Building Block   It has been proven long ago and in practice that the unity gain building block is the concept that is used in cascaded amplifier technology. Depending on the amplifier's output specifications, the concept is to allow the cable loss to equal the amplifier output level and vice versa. Essentially, the amplifier cable section calculates to unity gain; in other words, the cable loss equals the amplifier's output level. The cable loss is the highest at the upper frequencies, so therefore unity gain is calculated at the upper frequency limit. This concept is illustrated in Figure 1-3.

It should be obvious that the low-frequency end of the spectrum will arrive at the amplifier input at a much higher signal level than the high-frequency end since the low frequencies are attenuated a lot less by the cable section. The signal level at the low-frequency end of the spectrum has to be corrected by the internal equalizer network at the amplifier input section. Such equalizers come in a variety of values and are plug-ins, so the correct value can be selected for the specific cable section. Present day amplifiers still have selectable pads and equalizers to adjust the input signal level for each amplifier station.

The standard procedure is to measure the amplifier input signal level at both the low and high end (usually pilot frequencies) and then select the appropriate pad and equalizer to facilitate the setting of the output gain and slope. This technique will be treated in more detail later in the book. This procedure is known as amplifier or system balancing and is extremely important for the amplifier cascade to perform with optimum distortion and noise specifications.

Development of Coaxial Cables

Coaxial cable has undergone many developments over the years. The first type of coaxial cable had a braided copper wire shield and was covered by a plastic (polyethylene) jacket. Shielding effectiveness was poor because of the weaving of the braided shield, which allowed the cable to be quite flexible.

Figure 1-3
The unity gain
building block

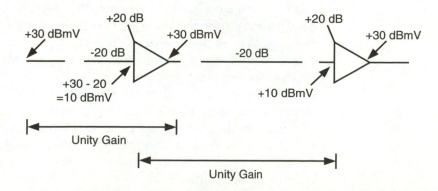

Connecting to such a shield, however, was a problem unto itself. The center conductor was often a copper wire or copper-plated steel wire. The insulation was often a harder polyethylene. Early coaxial cable was usually found on the surplus market left over from WWII. This type of cable had high attenuation, poor shielding, and was easily damaged by moisture. With the start of the cable television industry, cable manufacturers were quick to realize the need for more and better cable.

Solid Aluminum Sheath Cable   The solid aluminum-sheathed cable was introduced in the 0.412 and 0.500 in sizes with a nominal impedance of 75 ±2 ohms. The dielectric was either styrofoam or polyethylene foam and the center conductor was either solid copper or copper-clad aluminum. A polyvinyl chloride jacket was available and was recommended in coastal climates.

It was the solid aluminum sheath (tubing) that vastly improved the shielding effectiveness of the cable. In a great many areas, the bare aluminum cable worked well and became widely used. A whole series of connectors appeared on the market. Some were better than others and eventually the best became better as the cable television industry grew. The solid aluminum sheath cable was different to handle, that is, a lot stiffer, and could be easily kinked, thus destroying the transmission capability. Proper construction procedures, tools, and equipment were developed to correctly handle and install the cable in an expeditious manner. With the development of this cable, the cable television industry could extend its plants, further serving more subscribers with more channels with improved pictures.

Improvements in Dielectric   Various techniques were used to try and improve the loss versus frequency characteristics of coaxial cable. Since dry air has a velocity constant of 0.95, a cable with pure dry air as a dielectric should provide about the best attenuation versus frequency characteristic. Since some means has to be employed to keep the center conductor in the center, air just does not do the job. One cable manufacturer used polystyrene discs placed a few inches apart sealing in cells of air as well as supporting the center conductor in a concentric condition. Such cable did provide superior loss versus frequency characteristics, but it was difficult to manufacture since fusing the discs in place was a critical process. Some of this cable is still most likely in service and an improved version is available today.

Coaxial cable used today uses the polyfoam technique in which the plastic foam forms bubbles of air which control the dielectric constant. This cable has proven to be very rugged and maintains its characteristics over most climatic conditions.

Improved Attenuation and Loop Resistance The attenuation of cable as a function of frequency is a well-known characteristic to the cable television industry. Essentially, the loss or attenuation versus frequency is a logarithmic function, as shown in Figure 1-4. Most manufacturers publish a chart of attenuation versus frequency so system designers can calculate the cable span loss between amplifiers. The usual design technique is to calculate the span loss at the highest operating frequency that exhibits the greatest attenuation. This assures that each amplifier has the proper input and output signal levels at the upper frequency limit. The lower frequency end will have to use some equalization so the amplifiers are not overdriven. As far as cable performance, there have not been any large breakthroughs in cable loss characteristics, but several manufacturers have improved their manufacturing techniques, which have produced cable that is more rugged, has a lower return loss characteristic, and improved environmental performance.

Essentially, the loop resistance of coaxial cable depends on the conductivity of the center and outer conductors. Since the solid aluminum sheath cable has high conductivity for the shield, most of the loop resistance is attributed to the center conductor. Making the center conductor out of aluminum and cladding it with copper produces a lower resistance at the RF operating frequencies due to the well-known skin effect. However, the resistance at 60 Hz is appreciable enough to limit cable powering of the repeater amplifiers. By using a solid copper center conductor, the loop resistance of the cable span is decreased, resulting in possibly fewer 60-Hz power supplies.

Figure 1-4
Cable attenuation vs. frequency

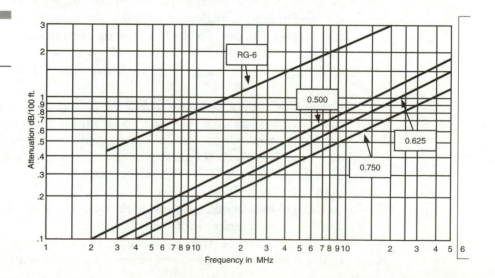

This fact has to be taken into account when working out the cost-per-mile figures. Also, the larger the cable size, the larger diameter center conductor, which results in lower loop resistance values. Again, lower loop resistance caused by either choice, solid-copper center conductor or larger cable size, can result in a fewer number of power supplies. Solid-copper center conductor cable and larger size cables are more expensive, so the cable designer has to weigh the pros and cons to arrive at the most cost-effective solution. It should be kept in mind that larger cable sizes can also result in greater spacing between amplifiers due to lower attenuation and hence could result in fewer amplifiers in a cable run.

Cable Jacketing, Flooding Compounds For many applications, bare aluminum-sheathed cable wears well. Typically, such applications would be in areas where there is no salt water environment. Coastal areas or industrial areas where air pollution couples with a wet climate, however, could cause a corrosive problem with aluminum. For such general corrosive environments, jacketed cable should be used.

In general, jacketed cable is more rugged simply because of the added strength of the jacket. Jacketed cable is heavier than unjacketed cable and, for the same size cable, is larger in diameter. This can limit the amount of cable length that can be placed on a reel, but this is usually not much of a problem because in practice the ends of a reel are often too short for use and are simply discarded. This is often referred to as shrinkage of cable stock.

Coaxial cable that is manufactured for underground installations is always jacketed and contains a flooding compound between the plastic (polyethylene) jacket and the aluminum sheath. If a break, cut, or gash occurs in the plastic, the flooding compound flows into the damaged area and seals it against moisture ingress. The use of polyethylene plastic in jacketing the cable, as well as the plastic bubbled foam used as the dielectric, has been responsible for some of the major improvements in the cable television industry.

Cable Connector Development and Standards Through the years, coaxial cable connectors have evolved from rudimentary to sophisticated, complex designs. One of the difficulties discovered early on was that many connectors had the problem of keeping a good electrical contact around the aluminum sheath. Loss of a complete connection around the circumference of the aluminum sheath caused RF energy to escape from the cable signal (called leakage) as well as electrical noise to enter the cable (called ingress), which in turn interfered with the cable signal.

Various techniques were used to assure good electrical connections for the center conductor and the surrounding aluminum sheath. Two types of connectors were developed. One was called the straight-through type, where the center conductor was carried through the connector to be fastened directly to the device being connected. The second type was referred to as the stinger type, where the center conductor entered a vise-like grip socket. Once the center conductor entered the connector, it could not be pulled out. This type of connector had a protruding pin called a stinger, which was connected to the center conductor. This pin was often trimmed to fit the connecting device.

The cable preparation for each of these connector types was different and had to be followed precisely to assure a proper fit. Several tool manufacturers made tools that would remove a portion of the jacket, core the cable dielectric, and cleave and clean the center conductors. Often the ends of the cable were prepared for a certain manufacturer's connector.

To obtain proper shielding characteristics, various methods were used. The cable was usually in some form of longitudinal stress due to its own weight as well as wind and vibration. Such tension could cause the cable to pull out of the connector, thus breaking the connection and interrupting service. The lashing operation alleviated most of the tension and the connector's grip on the cable sheath and center conductor prevented this from happening. To improve on the connector's grip on the aluminum cable sheath, a stainless steel sleeve was used to add strength to this grip. This sleeve was fit inside the aluminum sheath while the connectors wedge-type grip forced down on the aluminum against the sleeve, improving the connector's grip on the aluminum sheath.

At first, some stainless steel sleeves made for various cable sizes had to be installed first. The procedure was to remove some of the dielectric core, called coring, using a tool made for the cable size and specific connector and then inserting the sleeve. At present, most connector types have an integral sleeve placed within the connector that prevents a lost sleeve, an improper sleeve size, or simply forgetting to install the sleeve. Examples of a connector type are shown in Figure 1-5 and the preparation steps are displayed in

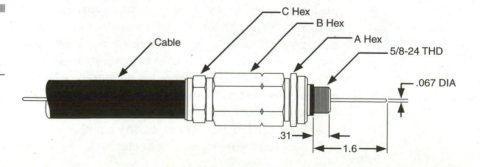

Figure 1-5

Pin type connector. Example of a cable connector

Figure 1-6
Cable connector
preparation

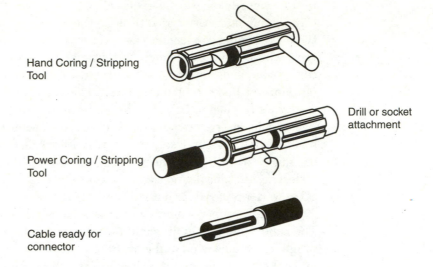

Hand Coring / Stripping
Tool

Power Coring / Stripping
Tool

Drill or socket
attachment

Cable ready for
connector

Figure 1-6. Proper installation of connectors of this type prevent signal leakage or noise and interference ingress as well as insure good weather proofing.

Since coaxial cable carries its own power, usually a 60-volt or 90-volt 60-Hz a.c., the cable connectors have to provide good current-carrying capacity, which is referred to as ampacity. Cable manufacturers cover the current carrying capacity of their cables along with other specifications in their catalogs and data books. The main thing is to select a cable type, an appropriate connector manufacturer, and then to order the accompanying tools to prepare the cable and install the connectors.

Construction

Few cable system operators install their own cable systems. Most systems are constructed by cable construction contractors who provide equipment and crews to do the job. These contractors often use subcontractors as specialists, such as splicing (connectorizing) contractors. The cable operator has to require that, in the case of a prime or principal contractor, approval be given to all subcontractors before work begins. A cable operator should also require that co-insurance clauses and hold-harmless clauses be part of the contract with the prime contractor.

Proper construction practices, work progress, and contract performance milestones are important to the contract's completion timetable. This work is sometimes performed by an independent consultant often required by the lending institution. Contract monitoring is extremely important to the cable operator during the construction phase of the project.

Aerial Cable System Improvements In many instances, an aerial plant is used by most cable operators since the power and telephone services are provided by such aerial pole line plants. Rural areas are often served by a utility pole line. Cable systems have to apply to the pole owners for permission to install the cable lines on the utility poles. These pole systems are either solely owned by the power utility or the telephone company, and in many instances the poles are owned on a 50-50 basis where the electric company and the telephone company each own half of the poles. Such pole ownership is known as joint-own or simply J-O poles.

Cable operators are leased space in a location usually one foot above the telephone line and 41 inches from the standard 120/220-volt power distribution service lines. The top section of each pole is reserved for high-voltage electric distribution service. This space is often termed the municipal space, reserved for municipal wires such as the wired fire alarm system.

If there is not any space for a cable system, as determined by a three-party survey, then the cable operator is required to pay the cost for either a new or taller pole. If proper space is available but the electric and telephone systems have crowded the pole, then the cable operator is required to pay the respective companies to rearrange their plant to make space available for the cable system. The process of applying for use, the survey of each pole applied for use, and the necessary work of rearranging the plant is known in the cable industry as make-ready. This is often a difficult and time-consuming process for the cable operators. No cable construction is permitted until this phase is completed and paid for. Once construction is permitted, work is allowed to proceed according to standard pole-line construction procedures, as written in the Telco Blue Book. Copies of this comprehensive construction manual are often made available to the cable operators by the telephone company to assure that the plant is properly built. For overhead or aerial plants, this is the way it has been done and is being done today.

Present-day aerial pole line plants are getting very crowded and larger poles have been required for upgraded and increased telephone and electric services. In many cases, the fire alarm-wired systems have been replaced by wireless systems that require a lot less maintenance and still have a high reliability. Removal of this plant has allowed more space for cable operators' expanded plants. The telephone companies in many instances are making additions to their plants using underground cable installations, and so are many cable operators.

Aerial plant equipment has improved throughout the years. Line amplifier housings have improved metals with upgraded RFI gaskets, as well as weatherproof gaskets. More attention is being paid to the characteristics of the many types of metals being used together in aerial pole line plants. Much has been learned in the past from experiences the telephone company has had with dissimilar metals, grounding, and bonding. Proper grounding procedures as directed in the Telco Blue Book are meticulously followed by cable construction contractors. The cable operators technical personnel should closely monitor the quality of construction of their new cable plant because they will have to live with it for years to come.

Progress in construction equipment has been forthcoming both in tooling and equipment. One of the most significant is the hydraulic lift trucks. Prices have become more reasonable since more truck makers are introducing their products to market. More used equipment is becoming available as well, so even smaller cable operators can afford to have at least one truck. Having a hydraulic lift truck available to a cable operator is important because a possibly dangerous situation can be handled safely using a bucket truck.

Underground Plant Development As mentioned earlier, placing a communications cable system underground is becoming more and more attractive. In heavily populated urban areas, underground facilities are often required by permit. Where underground ducts with connecting vaults are available, cable facilities could be placed. Leasing duct space from the telephone company, power company, or municipality by the cable system is often required. Often, this used to be the only choice, since there no utility poles were available. Cable companies operating under these conditions had to have the necessary equipment such as work area signs, barricades, vault ventilating blowers, and the like in order to perform underground plant maintenance properly.

In suburban and more rural areas, direct burial of coaxial cable is a choice when soil conditions allow for it. Sandy soil, soft clay, or gravel free of stones is best for direct burial of cables. Trenching or plowing of the cable provides a quick and efficient method of directly placing cable underground with a minimum of restoration. Where the soil contains a lot of rocks and stones, placing the cable in plastic duct is the best method. At first, trenching was used and the cable was placed in plastic duct, laid in the trench, and then the joints were cemented. Splices, electronic equipment, taps, and power supplies were placed in either metal vertical pedestals or horizontal plastic or metal vaults.

Today integral cable placed in flexible conduit is presently available for use for a variety of soil conditions and can be either plowed directly or placed in trenches. If several cables are to be installed at the same time, cable in conduit can be ordered in several color choices for easy identification later during any repair situations.

Aerial Lifts and Ladders   As the cable industry matured and improved, the equipment did as well. Originally, most construction was done by climbing personnel using belts and hooks. Many of the cable construction people got their training from either their former employer, such as the electric or telephone company, or the Army Signal Corps. Getting up and down a pole in a safe and expeditious manner was the name of the game. For several years, pole climbing and ladders tied off at the pole were commonly used by cable construction personnel. Then along came the lift vehicles. Some were simply electrically operated truck-mounted extension ladders and some were hydraulically operated jointed boom types. Both had their cost and operational features.

The vehicles used by the electric companies were usually insulated for certain required high-voltage levels and used insulated hydraulic control lines. Since cable operators were working above telephone lines and below power lines, they often used insulated trucks. The jointed hydraulic aerial basket truck was often the best choice because it provided a large amount of aerial positional articulation as well as the protection against shock hazards in case of accidental contact with electric lines. The equipment manufacturers were quick to realize the cable system applications and provided the industry with smaller and less expensive articulate, insulated lift vehicles. Cable operators were introduced to this equipment through local dealers and local trade shows. In short order, such equipment became available on the used market, making them available to cable operators with smaller budgets.

Some manufacturers of hydraulic lift trucks offered an accompanying variety of hydraulically operated tools. Drilling holes in poles, tightening the pole line hardware, and raising equipment all became quicker and easier with the use of such tools. Cable construction companies using such tooling could produce quality work faster than those using the more manual methods.

Some subscriber drop installers use ladders instead of the belt and climber methods. These ladders should be made of fiberglass (insulated), be placed at the pole, and be strapped to the pole to prevent accidental slipping. For personnel untrained in the use of belts and hooks, this is often the method of choice. A ladder is also required to connect the drop installation to the house, so a ladder is always a part of an installer's necessary equipment. Figure 1-7 and 1-8 show some of the aerial lift truck uses.

The proper safety practices using ladders, climbers, and lift vehicles are covered in many company safety manuals. Many cable operators have

Figure 1-7
Back-stringing strand
installation using a
bucket truck

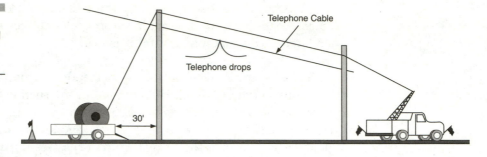

Figure 1-8
Back-stringing cable
installation using a
bucket truck

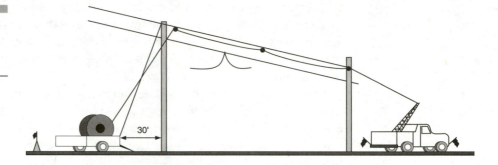

safety manuals covering their methods and equipment. The *Society of Cable Telecommunication Engineers* (SCTE) cover climbing and ladder use in their installer training manuals and lift truck manufacturers publish approved safe operation procedures.

Trucks, Trenches and Cable Plows   Many innovative techniques were tried when direct burial cable was introduced to the industry. Farmer's deep bottom plows were tried, which in some instances worked quite well, but the cable placed in such plowed trenches was not deep enough and stones were hard to find and remove. Equipment manufacturers soon introduced the chain trenchers, which were welcome to both the telephone and cable companies. Often a backfill "dozer" blade was mounted on the opposite end of the chain end for reversing the trenching direction, forcing the removed earth back into the trench after the cable was laid. Stones that could crush the cable had to be manually removed. When too many stones occurred, sand and gravel were used to cover the cable before the trench was filled. Trenching is still widely used today and, for many types of soil, it is the method of choice. Cable in conduit is often used today and is often faster because the backfill procedure is not critical.

As mentioned earlier, equipment manufacturers were quick to realize the potential advantages of opening a trench using the plow method. Today two types of plows are presently used to place cable underground: the vibratory plow and the stationary plow. Both types require little or no ground restoration. The vibratory plow is inserted into the earth and when the plow blade moves forward, it vibrates back and forth, forcing the earth to open in a slit. Behind the blade is the cable shute, which slides the cable to the bottom of the slit. The earth simply folds over, covering the cable. This type of plow is usually faster than the stationary plow and requires less draw-bar pull. Some vibration is transferred to the cable, however, which isn't sufficient to damage coaxial, telephone, or electric cable.

The stationary plow operates in essentially the same manner as the vibratory plow but does not vibrate. Thus, no vibration is transferred to the cable. This type of cable plow is required for the installation of buried fiber optical cable that cannot take the vibrations of the vibratory plow. The stationary plow blade is inserted into the earth and is pulled horizontally through the earth forcing a slit opening where the cable is inserted through a chute following the blade. This type of plow requires more draw-bar pull, which is provided by the drive engine and the tires or treads. Figure 1-9 displays these types of plows.

Another type of equipment used for urban cable system plants is the earth saw. This essentially is a huge circular saw with replaceable hardened teeth mounted on the edge. Rotating the saw with an engine can cut a three- to four-inch slit in either granite, concrete, or bituminous concrete road surfaces. Moving the rotating blade in a horizontal motion can cut a slit in hard surfaces at several feet per minute. In most cases, cables installed in such trenches are usually in some kind of conduit. Usually it is the integral cable conduit or steel pipe or PVC-type conduit.

For metropolitan installations, some kind of conduit is required by the public works department. Backfill and restoration is also in part specified by the municipal authorities. The cable conduit is often placed in soft sand at the bottom of the trench, which is then filled with a concrete slurry within a few inches to the top. The remainder is covered with the same material as the road or sidewalk surface. Cost-wise this is an expensive and noisy operation, but for urban-metropolitan installations, this is often the only method.

Metropolitan City Cable Plant   A metropolitan or city plant is much more expensive and complicated than the usual suburban, residential-type aerial and underground plants. Metropolitan areas often suffer from a variety of signal problems such as multipath and signal-level variations. These

Figure 1-9
Cable plowing
techniques

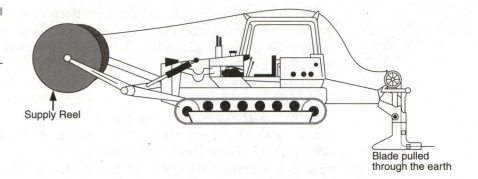

Supply Reel

Blade pulled
through the earth

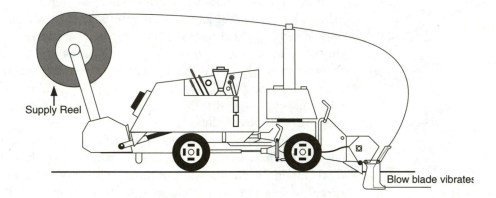

Supply Reel

Blow blade vibrates

metropolitan areas are usually covered by large apartment complexes where television service had previously been provided by a master antenna system and many apartment managers opted for cable television service when it became available. An apartment complex can make the usual cable television homes-per-mile requirement look good. Therefore, cable operators moved to the cities to provide service to these *multiple dwelling units* (MDU).

As stated earlier, a metropolitan plant is mostly an underground plant installed in either saw-cut trenches or in leased conduit space in the underground utility system. Cable operators using the underground utility plant have to use all the precautions and procedures as required by the conduit leasing agency. Safety barricades, traffic cones, and barricade tape are required to mark off the work area keeping the public away. Ventilating blowers are used to purge underground vaults of any unsafe gases and provide clean air for the maintenance technicians. Proper personnel training should be provided by the cable operators to ensure proper and safe working practices.

In many cases, the utility conduits pass under rivers and harbors in the transportation tunnels or across bridges and overpasses. Usually no access vaults are placed on bridges or in tunnels, but at each end. Thus, it is just the cables making the crossings. Cable amplifiers, couplers, and power supplies are installed in the vaults. Technicians making maintenance or troubleshooting checks have to gain access to the vaults to make the required tests. Cable service distribution is coupled from the vault system to the apartment buildings that contain the distribution amplifiers and subscriber taps.

An inter-building plant is either from top to bottom or bottom to top with the cables placed in utility shafts connecting to each floor level. At the various floors, a distribution panel is installed with drops connected to the apartments on each floor. In such urban areas, installation contractors often contract to install systems in new apartment complexes being built. Once such systems are built, the cable operators need properly trained personnel to perform connects, disconnects, and maintenance tests.

Changes in CATV Regulations, Requirements

Many changes have taken place in the area of governmental regulations. Most of the changes have been in the FCC regulations and are a concern for many cable operators. The other areas of change have been in the municipal or local regulations and for most cable operators no significant changes have taken place. Both of these types of regulations will be discussed.

Federal Regulations, Communications Act of 1934

The FCC regulations pertaining to cable television operations have increased in number and complexity over the years. Early start-up systems essentially had no type of regulations and at that time the commission could see no need for any controls. When cable systems improved and the expanded services were offering distant imported television stations, the must-carry rules were formulated and are still with us today.

Additions, Rules for CATV   When cable systems started to use the so-called mid-band, between channels 6 and 7, for added services, the leakage regulations were imposed by the FCC. The cable industry worked with the commission in developing the *Cumulative Leak Index* (CLI) concept, which is in use today. These rules require leakage measurements of signal levels from cable systems using a signal level meter and a test antenna. The measurements are then used to calculate the signal in microvolts per meter and, by using a formula with the number of leaks measured, the CLI for the system can be calculated. This procedure and an example are given in Appendix A.

Another area of concern that the FCC has for cable television systems is in the Proof of Performance Tests and how and when the tests are made. As the cable television industry has progressed, the commission has become more concerned with the signal quality of the product. Systems now have to test for base-band video signal quality. Proof of Performance regulations are listed in Appendix B.

Cable Television System Congressional Bill   Probably the most important regulation or standard adopted by the FCC is the recently accepted HDTV standard. The work performed by the Grand Alliance and the FCC has produced the standard for HDTV. The Grand Alliance is made up of experts from AT&T, the *General Instrument Corporation* (GI), Zenith, Thomson, Philips, Sarnoff Labs, and the Massachusetts Institute of Technology. Its purpose is to recommend HDTV standards and methods of modulation. The FCC created the *Advanced Television Systems Committee* (ATSC) to complete the work of the Grand Alliance from an official standpoint. The ATSC-SI-issued Document A/56 outlines and addresses issues for subscriber devices and services for such digital programming. Since its formation, Cable Labs in Colorado has been very active and supportive in working with ATSC. With all the work that the industry and the FCC has been doing to get HDTV and digital television up and running, many questions remain unanswered at this time. This means more work is to be done to resolve the must-carry questions, and the restrictions and controls applying to the cable television industry.

In December of 1996, the FCC gave the OK for broadcasters to begin testing the new digital television transmission. This is the standard for the broadcasting industry to follow for transmitting HDTV, video, and audio services to the public. Since new television receivers will be needed to receive HDTV signals, a time frame for conversion has been set. After 2003, all television broadcasters will be required to be transmitting digitally in either HDTV or NTSC video formats. Modulation for broadcasters will be 8-VSB, while cable systems will use the QAM-16, 64, or 256 digital formats. These methods of television transmission will be discussed in Chapter 4, "Digital Technology & Cable System Applications."

Major Effects on The Industry   From time to time, Congress has passed legislation specifying cable television regulations and will continue to do so. The cable television industry has evolved from the simple systems of the early 1950s and 1960s to a multibillion-dollar telecommunications industry. Still, many small simple systems that are long since debt-free continue to provide one-way multichannel service to subscribers. The normal course of events for the industry was to consolidate assets by mergers, thus forming several giant corporations. Such corporations activated the return path, expanded the signal bandwidth in the forward direction, installed fiber-optic plants, and activated pay-per-view services. Some even instituted telephone and data services to become an all-service telecommunication provider. These are the issues the FCC has to contend with as the telecommunications industry grows.

The FCC issued its second report and order on December 31, 1998 that requires cable television systems to become a part of the *Emergency Alert System* (EAS). This order requires equipment to be installed in cable system hub/head-ends so that in times of emergency an alert tone and message crawl is placed across all cable channels. Broadcasters also have this capability and are responsible to provide the emergency alert to people receiving just off-air reception. Several manufacturers also offer equipment and installation guidelines to cable systems. This is a good and useful service to cable subscribers and could provide appropriate warnings of hurricanes, tornados, and disasters.

Local Regulations

Most of the local regulations are written into the licensing/franchising agreements between the municipality and the cable operator. These regulations or requirements, as the case may be, address local programming issues, cable channels allocated to community use, and the equipment to produce such programs. Regulations regarding the cable operators' right to work on the public ways is also part of the licensing agreement. Some more urban communities have more extensive rules and regulations concerning work along the public ways. The department of public works usually is the community administrator of these rules and regulations. The local telephone company and the electric company also fall under these same rules. Essentially, the rules and regulations are concerned with public safety and the protection of public property.

Permits for Construction   The cable television license or franchise for a municipality or community constitutes permission to work along the public highways and byways. This license or franchise has to be in place legally before construction can commence. Proper insurance coverage for public liability and property damage is a requirement due at the signing of the license. Often safety issues are written into such agreements. Also at the time of licensing, the pole attachment agreement with the pole owners, usually the telephone and power companies, has to be in place. Some communities themselves are the local pole owners. Communities that usually own their own poles also are the municipal electric power provider. They in turn lease space on the poles for telephone and cable television systems.

Inspection and Control   After construction of a cable television system is completed, the pole owners usually inspect the plant to ascertain if it is in compliance with the agreement. This is important to all parties concerned to resolve any issues or problems with the new cable television system. In many cases, the local department of public works will want an inspection of the construction along the public streets. A representative of the cable television company and that of the town or county will walk or drive through the various streets to ascertain that waste removal and restoration work has been completed in a satisfactory fashion. Once these inspections are completed, all concerned parties are assured that construction and cleanup is over and settled to everyone's satisfaction.

Commercial and Consumer Electronic Development

One very important factor facing the cable television operator is the subscriber equipment, mainly the television set itself. Unfortunately, this device is beyond the control of the cable system and belongs in the consumer electronic manufacturing domain. As most of us in the cable television business know, relations with the television set manufacturers have not been smooth throughout the years.

The organization representing the consumer electronics industry is the *Consumer Electronic Industry Association* (CEIA) or the *Consumer Electronic Manufacturers Association* (CEMA). This association evolved from the *Radio Manufacturers Association* (RMA). The CEIA represents the manufacturers of consumer electronics equipment, most of which consists of home entertainment equipment.

It is well known that the general public spends large amounts of money on television, stereo, telephone, and computer equipment. One of the difficult areas of contention between cable television operators and the consumer electronics industry is that of cable-ready television sets. With digital television just around the corner, there is a great need for CEMA, the broadcasters, and the cable television operators to settle their differences and get down to the business of resolving the digital television issues. It is indeed in the best interests of the general consumer public that straightforward and easy-to-integrate equipment be available to interface the television set and the cable system.

Cable systems have made the converter box available to subscribers, with some integral form of decoder/descrambler for the premium (extracost) channels. This box also has a companion remote control that enables the converter to select a cable channel and present it to the TV set on a fixed channel, usually television channel 3 or 4. The subscriber now has to keep his TV set tuned to the converter channel and the program channel is selected by the converter remote control. This renders the TV remote control almost useless and having two remote controls is often a confusion factor. Television sets that are advertised as cable-ready are able to tune to the usual cable television channels using the TV set's remote control. For nonpremium services, this works best, but when premium services are desired, the converter-decoder set-top device is needed with its own remote.

Some television receivers have what is called the smart remote which in many instances is programmed to operate the converter as well as the VCR and the remote. Now only one all-purpose remote is needed. Many manufacturers of VCR's also have programmable remote controls. Many home entertainment systems containing the TV set, VCR, a surround sound system and a video disc system may have as many as 5 or 6 remote controls. This can result in confusion. A few examples of subscriber equipment connected to the cable system are shown in Figure 1-10.

Figure 1-10

Examples of cable system to subscriber equipment

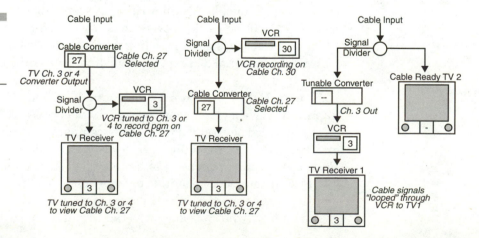

Rise of Digital Techniques Digital-formatted television is definitely coming and soon. In the past several years, much experimentation has taken place with advanced high-performance television methods. The Japanese experimented with an analog system that showed significant improvement. However, with the efforts of several manufacturers, the FCC and Cable Labs, as well as the Society of Motion Picture and Television Engineers, the digital method *Motion Picture Experts Group* (MPEG) II was approved and adopted. This MPEG II method carries the digital video signal in a compressed format along with CD-like *multi-television sound* (MTS) quality. The digital numbers forming the bitstream activate pixels of the proper color on the display screen. In compressed form, only the changes in picture content are transmitted, resulting in a decrease in the overall bit rate required to provide a picture frame.

Using digital technology, the encryption or encoding needed for premium programming can be built into the digital information stream. For an aspect ratio of 4:3 (NTSC format) and with present video picture quality, the video signal in compressed format requires less channel bandwidth than the present 6-MHz television band. Thus, four equivalent television channels with NTSC present quality standards can be carried in the six-MHz band with MPEG II video compression technology. This implies that each television broadcast station can gain three more equivalent channels, along with three times the advertising slots, by converting to the MPEG II digital compression standards.

Broadcast stations desiring to carry the *High Definition Television* (HDTV) standard signal form will require full use of the 6-MHz band for both video and sound. The HDTV standard uses a 16:9 aspect ratio and more pixels per picture, resulting in near-35mm motion picture quality. Several broadcast stations in the three major networks are preparing their transmission facilities for digital signals. At this writing, it seems that the VSB-8 method of modulation will be used for over-the-air broadcasts and will be covered in Chapter 4. Cable systems are opting for *Quadrature Amplitude Modulation* (QAM) 16, 64 or QAM 256 digital modulation for over-the-cable transmission.

ATV and HDTV Standards As stated earlier, the initial trial into improved television transmission quality was called *Advanced Television* (ATV) in which several manufacturing companies experimented with several methods. The resulting push for some sort of standards resulted in General Instrument Corporation's research on digital television technology. The FCC appointed a committee to investigate all of the proposed methods, and the acceptance of MPEG II digital video compression was the obvious choice. At first, what was known as the Grand Alliance formed by industry leaders, as mentioned previously, studied the ATV and HDTV proposals. The FCC-appointed committee was formed from this Alliance. As mentioned before, the present MPEG II digital compression standard was approved. The FCC has given the so-called green light to trials in 1998 and has allocated frequency bands to carry the new television digital format. The year 2003 is the target for full digital television broadcasting. Presently, several broadcasters are readying their transmission facilities for digital transmission trials and testing.

Wide Screen TV, VCRs, Video Disc Players Involvement   At present, digital television receivers are not on the market. Most consumers seem to think some form of conversion box or converter will become available to convert the digitally transmitted format to standard NTSC format that is compatible with present-day TV sets. Although this seems to be a reasonable expectation, there appears to be no such equipment forthcoming. If broadcasters opt for a 4:3 aspect ratio digital television or an HDTV 16:9 aspect ratio digital television, it begs the question, does this mean two types of television sets will be required? And if 4:3 ratio pictures are shown on a 16:9 wide screen, will HDTV TV sets be accepted? Since there are no direct-view cathode ray display tubes in the 16:9 format, what type of display method will be used? Some people believe that rear-screen projection type TV sets will be the first, followed by the yet-to-be-developed CCD color flat panel display. Clearly with the transmission of HDTV coming in the near future, television sets capable of receiving and displaying the new format should be ready approximately around the same time. Failure to do so will jeopardize the successful introduction of the service.

The 16:9 aspect ratio will probably take some getting used to in the living/family room setting. A comparison of 4:3 and 16:9 screens for nearly equivalent viewing areas is shown in Figure 1-11. For large screen areas, the 16:9 format may be too wide for many wall spaces. This could cause some problems with acceptance. The flat panel display, wall mounted, will probably be the best method for HDTV wide-screen viewing.

Figure 1-11
Television screen size
for 4:3 and 16:9
aspect ratios

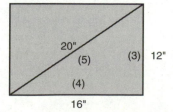

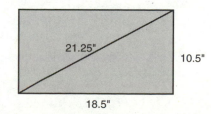

At present, some VCRs use an internal digital video format, and the video disc player also uses digitally recorded discs. Presentation to current TV sets would require the conversion to standard NTSC video/audio for use by the NTSC TV sets. When digital TV sets become available, then the digital output signals from the digital tape and disc machines will connect directly to the TV set. The internal RF modulators in digital VCRs and disc players, as well as the RF channel selector in the TV set, should be circumvented so the digital signals will avoid any added noise.

The consumer public anxiously awaits the arrival of digital television and its promised improvement in picture and sound quality. The prospects of the HDTV 16:9 aspect ratio in large screen format being accepted by the television viewing public, however, is anybody's guess.

Tree/Branch Cable Systems and Network Topology

Introduction

The basic idea of the cable television system was that the central antenna system would be shared among connected subscribers willing to pay a monthly fee. The head-end, as it was known and still is, was the source of the signals making up the basic service. The head-end was usually located on a hilltop, tall building, water tank, or steel tower located out of town away from the subscriber homes. The basic off-air received television signals were amplified, filtered, adjusted in level, and combined and delivered by the cable amplifier cascade to subscribers. This elementary topology was known as the tree-branch cable system. The main cable was the trunk system from the head-end to the town where the cable branched down the streets to the subscriber's homes.

A One-way System, Two-way Considerations

The basic tree-branch system consisted of cable-amplifier sections feeding signals one way from the receiving head-end tower site to subscribers. Many early cable operators were aware of possible two-way operations by splitting the service signal band between downstream and upstream service and installing upstream amplifiers. Such early thinking was basically wishful thinking mainly because of system problems such as temperature stability, signal level control, and amplifier noise and distortion. Basically, early systems were one-way systems and even today many systems are still essentially one-way.

Product Delivery, Many Channels   The bread and butter of cable systems operation was, and in many cases still is, the delivery of many cable television channels containing a variety of program choices to subscribers. When more channels were offered and cable system costs increased, subscriber rates were subsequently increased. The concept of program tiering allowed subscribers a choice of programming packages such as news, sports, and movies. The source of these programs came about with the introduction of satellite technology. Presently, the number of program channels available from various satellite systems is enormous.

The two problems cable system operators have in today's market are (1) adequate system bandwidth needed for an increase in channels and (2) making the decision as to what programming will sell best. To repeat, the bread and butter of cable system operations is the delivery of television programming to subscribers. It was thought at one time that auxiliary services such as electric meter reading and alarm systems would support two-way cable operations. The cost of the reverse system activation is a significant expense compared to projected revenues for the services.

Service Area Specifications The problem of identifying the service area of a cable television system is a difficult one. First and foremost is the fact that the centers of residential population are the criteria for successful subscriber penetration. Unfortunately, the licensing or franchising of municipal authorities may impose requirements that non-productive areas still be served. Also, future areas of residential expansion may not be readily identified, thus necessitating future plant expansions.

Most cable system operators avoided building cable plants in industrial and commercial areas. Thus, many cable operators have had to build more plants to provide information services to these areas. The location of the subscriber office area, the technical maintenance area, and the head-end location can also be factors in planning and specifying the service area.

Essentially, the service area has to be specified before financing and construction plans can commence. Thus, as the saying says, do the homework carefully and diligently. Technical people as well as managerial people should be involved in this decision-making process. By estimating the miles of the plant, the feeder-to-trunk ratio, and the number of program channels, the cost per mile of plant and the head-end costs can be projected. Now the basic revenue can be projected, producing a cash-flow or financial pro-forma to offset debt and support the venture. This process is complicated and has to be carefully completed before any construction schedule can be made.

Expansion of Feeder Criteria The proper plant design required to serve the more heavily populated area of many communities should not be so trunk-restrictive that businesses will box themselves in by making it difficult to extend the distribution cables to new areas. In short, a system does not want to run out of trunk that can restrict extending the feeder plant. It must be remembered it is the feeder or distribution plant that connects the cable system to subscribers. The trunk is considered as the transportation system bringing the service to the distribution plant.

By now, it should be obvious that the feeder/distribution plant makes the money, while the trunk cable plant is a necessary cost. The extension of a feeder plant adds more subscribers, thus improving the cash flow. Since the cost of extending a feeder plant is significantly less than the trunk plant costs, a careful analysis has to be made when extending a plant. Often a

Figure 2-1

Trunk extension
required to extend
service area

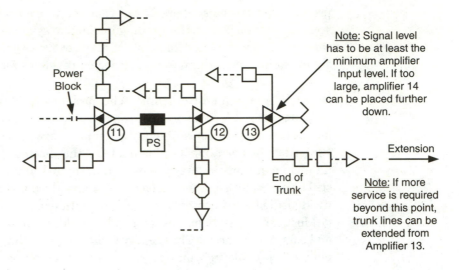

Note: Signal level
has to be at least the
minimum amplifier
input level. If too
large, amplifier 14
can be placed further
down.

Note: If more
service is required
beyond this point,
trunk lines can be
extended from
Amplifier 13.

Figure 2-2

Added trunk requires
added system power
supplies

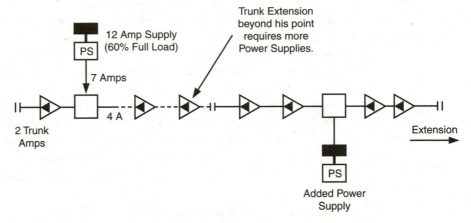

significant extension of the feeder plant means adding some more trunk plants. This is shown in Figure 2-1, where feeder plant amplifier cascades are limited to two amplifiers in a cascade off the bridger trunk amplifier.

Adding more cable amplifiers to a system also means that system powering may have to be restructured with the addition of more power supplies and resegmenting the powering system. Repowering such plant additions is shown in Figure 2-2.

Television Receivers

The main receiving device of a cable television system is the subscriber's television receiver. These receivers can range from new to almost museum-piece television sets. However, since television sets have an average life of about 10 years, it is fairly safe to assume that most subscriber sets range from new to 10 years old. The median age range is estimated to be about four or five years old. Also, almost all of the television sets are color TV sets.

Since cable systems carry the service in standard NTSC VSB format, the television receives the signals and makes the channel selection by frequency-synthesized tuners, the same as if it were an off-air TV station. Since cable system channels can be offset in channel frequencies in the HRC or IRC format, television sets made in the last 10 to 12 years have a switch often appearing on the rear cover or selected by the remote control labeled Cable 1 or Cable 2. This switch controls the action of the channel selector to tune to the desired channel settings.

The present-day television set is devoid of horizontal and sometimes vertical hold controls due to vastly improved circuitry that locks the picture synchronization. Automatic chroma controls and contrast controls keep the pictures at the same color and brightness. The remote control system can usually address all picture quality controls, channel selection, and video source selection, making most of us the so-called couch potatoes.

Integrated circuit technology, circuit board construction, and modular tuning, along with the remote control system, have reduced the cost of television sets drastically. Still the picture tube or the projection system determines the cost of the receiver. All of the electronic systems of a number of present-day television sets are valued at $40 to $50. The cabinet and screen/tube system results in most of the cost of a TV set. Since the subscriber's television set is what produces the cable system's picture, it is most important that cable operators make sure the cable installer makes the proper connections to the TV set and instructs and advises the subscriber in getting the most out of the service.

Channel Tuning As stated earlier, the television set has to tune precisely to the cable channels for best signal and sound quality. The noise characteristics of the television set tuner are most critical. The noise figures of early TV sets were quite poor and the UHF portion even worse. As the TV sets improved, the noise figures of the channel-selecting tuners did as well. The IF bandwidth characteristics improved also, which in turn gave better pictures with improvements in the quality of TV receiver design. The picture resolution improved from about 260 lines to about 380 to 400. With the arrival of HDTV, picture resolution will indeed improve again.

Since HDTV will be digital, as proposed and scheduled in the next century, the consumer electronic manufacturers are going to be busy designing, developing, and testing TV sets for this new service. The broadcasting companies are busy getting new studio and transmitting equipment capable of handling the digital signals, readying the public for HDTV service where no TV HDTV sets exist. The trials of off-air digital television services will be taking place soon. Then maybe the problem areas will be identified. As yet no signal strength parameters, signal-to-noise specifications, or the tolerance-to-multipath reception information has resulted. In short, no one knows if digital television broadcast stations can be received with rabbit-ear antennas or will be back to roof-top antennas.

As far as the cable operator is concerned, questions remain. First, will cable head-ends receive digitally modulated signals from a tower-mounted antenna? Second, will cable systems convert some of the signals to other modulation schemes or to NTSC VSB format? Third, what will be the must-carry rules as set forth by the FCC for broadcast stations transmitting multiple channel digital signals? And the list seems to go on and on.

Channel Converters The channel converter either rented or purchased from the cable operator has traditionally been required to provide cable television channels unable to be selected by standard TV sets. For some early cable systems that placed added programming channels in the so-called mid-band, a simple UHF-up converter was used. This device simply converted the mid-band channels that were spaced from the FM band to VHF Channel 7 to the UHF band (470 to 890 MHz). This block converter, as it was known, connected between the cable and the TV set's UHF terminal, as shown in Figure 2-3.

Since there were nine equivalent 6-MHz television channels spaced in the mid-band, the nine channels, in addition to the 12 VHF channels, resulted in what was known as the 21-channel cable system. At this point in time, available trunk and distribution cables had significant loss at UHF frequencies. Available amplifiers had upper frequency limits of 220 to 300 MHz. The reception of UHF broadcast stations was carried on the cable systems by converting each one of them to one of these nine mid-band cable channels at the head-end. The blockup converter at the subscriber's television set would then convert them back from mid-band to UHF.

Back in the 1970s, the 21-channel cable system offered a decent selection of programming with better reception than most indoor or outdoor antennas. The success of 21-channel systems encouraged the expansion of cable operators to increase the upper frequency limit to 300 MHz. Thus, another 14 television channels were added, making the 35 channel system. This increase in channels beyond the frequency selection capabilities of the television sets of the day made the tunable or frequency-agile set-top converter

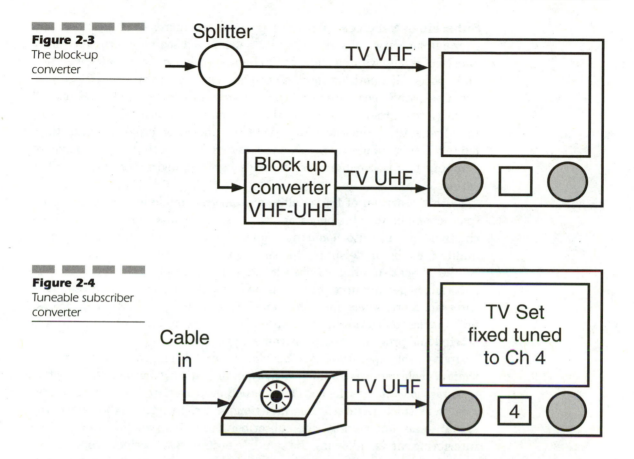

Figure 2-3
The block-up converter

Figure 2-4
Tuneable subscriber converter

necessary. The method of connecting this device between the cable and the television set is shown in Figure 2-4.

Converters of this type had to have good noise specifications, or low noise, simply because the cable signal has to go through the so-called front end of the converter and then through the front end of the television set, resulting in noise contributions from both sources. These first generation tunable converters required the subscriber to select the desired program on the converter, which translated to a fixed channel (usually channel 3 or 4). This situation essentially defeated the use of the set's remote control. Thus, the first big problem with subscriber devices resulted. A natural progression to increase the upper frequency limit beyond 300 MHz was forthcoming. Thus, the evolution from 52 channels at 400 MHz to 62 channels at 450 MHz to current state of the art systems offering more than 150 channels resulted.

Subscriber Services   With the tremendous increase in cable system channel capacity, subscriber service selections became enormous. Many subscribers only wanted a few basic channels covering local news, weather, and sports with nothing more. Therefore, cable operators tiered the program selections according to their subscribers' viewing habits. Since all cable system programs were on the system, it was natural to use a filter installed at the service drop to prevent the undesired program bands to be fed to the subscribers' home. This was referred, in the industry, as negative trapping; that is, the unordered services were trapped out of the subscriber service drop.

With the arrival of the satellite system that made available premium (pay) channels to cable operators, the negative traps could prevent these channels getting into homes not subscribing to these channels. System audits could be made simply by passing the aerial plant taps and looking for the presence or absence of the traps. This method was easily defeated by subscribers getting up to the tap usually by use of a ladder and removing the trap. More determined cheaters replaced the trap filter with a look-alike device to fool the cable operator. Hence, the pilfering of cable services started and program security became an issue.

Today cable operators have developed various programs to control the theft of cable services. From negative signal-trapping methods, the positive trap method was developed. This method employed an interfering carrier system in which an interfering carrier signal was injected in the band of the signal to be protected. At the subscriber tap, a trap would filter out the interfering carrier, allowing the signal to enter the subscriber drop without the interfering signal so the subscriber could receive the premium program. This method meant that only premium (pay) program subscribers would need this trap. The ratio of premium subscribers to non-premium subscribers was an important factor for making the economic decision for the best method. In short, a system that does not have many pay program subscribers needs a greater number of negative traps than positive-type traps.

It must be remembered that the programming choices are the stock and trade of the cable television industry. In today's market where the computer Internet connection is a desired service, many telephone system service providers have difficulty in delivering large data files because of the limiting transfer speeds or low bit rates. A cable system's bandwidth can allow for faster downstream bit rates, but upstream ordering capabilities are in many cases not available. For cable systems operating in the present computer environment, the expansion toward two services is a must. Bidirectional plants with sufficient bandwidth are necessary for cable operators to become players in the telecommunications field.

The Head-end

The cable system head-end is the source for all the programs offered on a system. It is also the main source of control, often referred to the root of the trunk/branch-tree network. From this point, signals are delivered to subscribers via the trunk/distribution cable system known as the limbs and branches of the tree network topology. The head-end system can vary from one central point to a system of other source points feeding a common source node. Head-end signal system networks can range from simple to several locations connected by a trunk cable system. Some head-end network topology is shown in Figure 2-5.

Antennas and Off-Air Reception

One very important source of signals most subscribers watch and depend on is the local television broadcast stations. These stations are the local

Figure 2-5
Examples of
head/end-hub
configuration

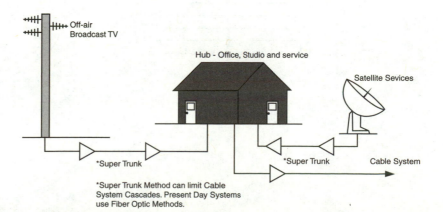

network affiliate in the standard VHF band plus the independents and the public broadcast stations. Local news, weather, and sports are usually favored programming. After all, cable television got its start carrying the off-air broadcast stations to distant areas or areas where the signal was of poor quality. Usually cable companies receive the off-air broadcast stations from tower-mounted antennas feeding the signal to amplifiers or signal processors in the head-end electronic equipment building. Usually this building was placed near the foot of the tower. An ideal situation would be to have the satellite receiving antennas and equipment at this location as well.

Antenna Types Log periodic antennas and Yagi antennas are the two most common types for VHF reception. Figure 2-6a–b shows these two types of antennas. The log periodic is a bit more broad-band and hence can be used for two or three adjacent channels, providing the stations are in the same pointing direction. This antenna has good gain, directivity, and high front-to-back ratio characteristics and is usually the antenna of choice for moderate to distant television stations.

For UHF TV station applications, the bow-tie corner reflector can be used for stations not too far away and for distant stations either the wire type or solid-aluminum parabolic antenna is usually the best choice. These antenna types are shown in Figure 2-7.

Placement on the antenna tower or mast is important and, unfortunately in many instances, is overlooked by some cable systems. The close

Figure 2-6a
Yagi pattern

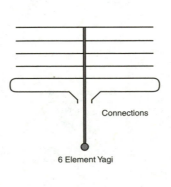

Connections

6 Element Yagi

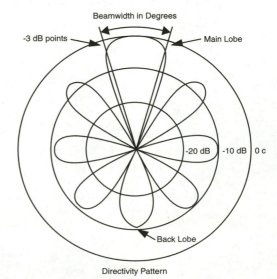

Beamwidth in Degrees

-3 dB points

Main Lobe

-20 dB -10 dB 0 c

Back Lobe

Directivity Pattern

Figure 2-6b

Log perodic antenna

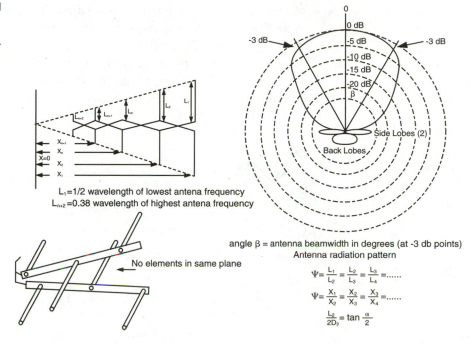

L₁=1/2 wavelength of lowest antena frequency
Lₙ₊₂=0.38 wavelength of highest antena frequency

No elements in same plane

angle β = antenna beamwidth in degrees (at -3 db points)
Antenna radiation pattern

$$\Psi = \frac{L_1}{L_2} = \frac{L_2}{L_3} = \frac{L_3}{L_4} = \ldots\ldots$$

$$\Psi = \frac{X_1}{X_2} = \frac{X_2}{X_3} = \frac{X_3}{X_4} = \ldots\ldots$$

$$\frac{L_2}{2D_2} = \tan\frac{\alpha}{2}$$

Figure 2-7

Corner reflector and parabolic antennas

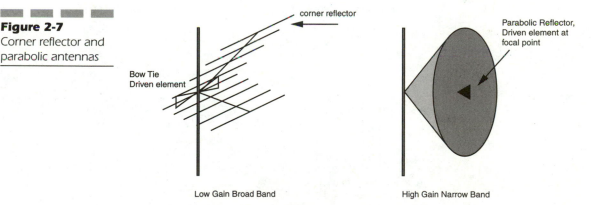

corner reflector

Bow Tie
Driven element

Low Gain Broad Band

Parabolic Reflector,
Driven element at
focal point

High Gain Narrow Band

proximity of antennas can interfere with the antenna beam patterns of each, and proper spacing, depending on the wavelengths of the antennas, is required to overcome this problem.

Placement on Tower   Placement on the tower between VHF antennas is often the main consideration. Since UHF stations operate at shorter wavelengths, the separation between antennas can be closer. Also the location of antennas depends on the pointing direction. For example, antennas placed at directions on opposite tower legs can be spaced vertically closer. This case is shown in Figure 2-8.

Notice that antennas pointing in a direction 180 degrees from each other can be placed on different tower legs at the same level. When working on the antenna placement problems on a tower, the result will be the necessary amount of tower height needed for the required antennas. Once this mea-

Figure 2-8
Antenna mounting
location on tower

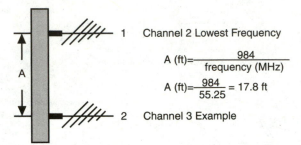

1 Channel 2 Lowest Frequency

$$A \text{ (ft)} = \frac{984}{\text{frequency (MHz)}}$$

$$A \text{ (ft)} = \frac{984}{55.25} = 17.8 \text{ ft}$$

2 Channel 3 Example

1. The antenna for the weakest station should occupy the topmost section of the tower.

2. If the receiver signals are equal, place the highest frequency (smallest) on top so as not to weigh the top unnecessarily.

3. Follow the procedures outlined for proper minimum allowed spacing.

Vertical Separation of Antennas
(a)

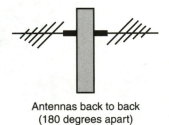

Antennas back to back
(180 degrees apart)
(b)

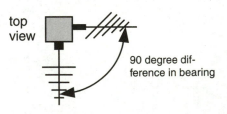

90 degree difference in bearing

$$A_{min} = A \left(1 - \frac{0}{180}\right)$$

$$= 17.8 \left(1 - \frac{90}{180}\right)$$

$$= 17.8 \left(1 - \frac{1}{2}\right) = 17.8 \times 1/2$$

$$= 8.9 \text{ ft (or 1/2 what it was before)}$$

Antennas 90 degrees apart)
(c)

surement is known, the next issue is the amount of added height needed for proper signal strength. The usual method involves an on-site signal survey, which involves some expense.

First and foremost is obtaining the broadcast station signal contour maps, which are often available from either the antenna manufacturers or from a consulting firm specializing in performing such on-site signal surveys. Data taken from these maps can tell whether a full signal survey is necessary. In many instances, the cable system technical staff can simply go to the site and erect a temporary wide-band rooftop antenna and take signal-level measurements and picture-quality measurements with a signal-level meter and a television set. If any interference is evident, further signal tests with a spectrum analyzer may be necessary.

Pre-amplification If some of the desired stations may be quite weak due to the distance from the station, a pre-amplifier may solve the problem. In some instances, the pre-amp can be placed in the head-end building before the signal processor, or if the signal is extremely weak, the pre-amp can be placed on the tower at the antenna. Tower-mounted pre-amps usually are fed their power through the coaxial cable downlead. Care must be taken to maintain good coaxial connections with no chance of moisture ingress so corrosion problems can be prevented. The signal-to-noise ratio is the most important parameter for the pre-amp placement problem and is shown in an example in Figure 2-9.

For UHF stations, the downlead loss can become a significant parameter for particularly tall towers or where the head-end off-air building is separated a long distance from the tower. If this loss is significant, a tower-mounted converter can be used that converts the UHF signal to a VHF signal, which now has a lower cable loss. Often these converters contain a pre-amp that improves the signal-to-noise ratio. An example comparing the results of the use of a converter and the results without one is shown in Figure 2-10.

Cable systems requiring tower-mounted converters and/or pre-amps are usually at remote areas far from the broadcast stations. For metropolitan cable systems, often a tall building's roof can be used to circumvent the multipath reception problem. For such systems, signal level is not a problem so small, more elementary antennas are sufficient. Again, an on-site signal survey will tell the story and indicate the needed equipment.

Tower Types Most cable systems use some form of tower or antenna mast to support the off-air antennas. Therefore, some space should be devoted to the discussion of towers and tower maintenance. Towers over the

Figure 2-9

Pre-amp location at
head end

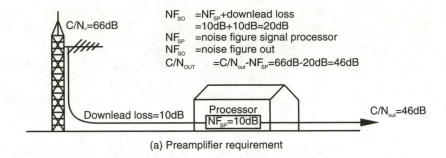

(a) Preamplifier requirement

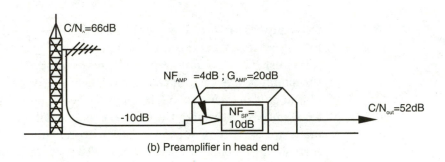

(b) Preamplifier in head end

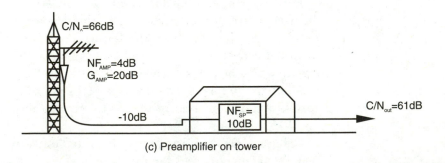

(c) Preamplifier on tower

usual 200-foot height are considered obstructions and are a hazard to airplane traffic. Proper application should be made to the obstruction department of the regional *Federal Aviation Administration* (FAA) office. Information pertaining to the location of the office can usually be found in the telephone book under U.S. Government or by calling the local regional airport. The FAA office will provide the necessary application forms and procedures to apply for permission to construct. Any required lighting and painting requirements will be specified by the FAA.

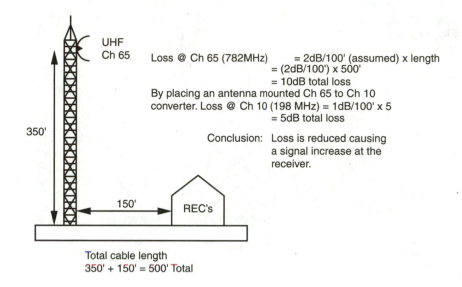

Figure 2-10
Tower mounted
amplifier converter
application and
signal level
comparison

UHF
Ch 65

Loss @ Ch 65 (782MHz) = 2dB/100' (assumed) x length
= (2dB/100') x 500'
= 10dB total loss
By placing an antenna mounted Ch 65 to Ch 10
converter. Loss @ Ch 10 (198 MHz) = 1dB/100' x 5
= 5dB total loss

Conclusion: Loss is reduced causing
a signal increase at the
receiver.

350'

150' REC's

Total cable length
350' + 150' = 500' Total

Normally, towers over 200 feet and less than 300 feet usually require a flashing red beacon light at the top, a set of three fixed side lights at the halfway mark, and orange and white paint alternated at usually 20-foot intervals. Taller towers may require a flashing beacon at its center in addition to the top beacon. Also, two sets of fixed side lights between the flashing beacons are required at its center.

Towers fall into two main categories: self-supporting and guyed towers. Self-supporting towers are usually three-legged, but some older ones may be four-legged. They are generally more expensive but use less land than guyed towers. Large self-supporting towers are usually too expensive for most cable television applications. Guyed towers are more common for cable applications and range in height from 100 feet to 300 feet. Higher towers require more land for the guy wires and often use three inner guys and three outer guys. A comparison between the free-standing tower and a guyed tower is shown in Figure 2-11, showing the relationship between the height and land requirements.

Guyed towers are usually triangular and may have either a fixed based mount or a tapered based mount, as shown in Figure 2-12. The difference is in the distribution of the bending moments or forces on the tower legs. For shorter guyed triangular towers, the fixed base towers are often used. For solid rod towers, which are usually a lot heavier, use the tapered mount.

Manufacturers of towers usually have a customer engineering service department available to assist the customer in specifying an appropriate tower. Tower installation crews can often be arranged through the

■■ ■■ ■■ ■■
Figure 2-11
Head-end towers

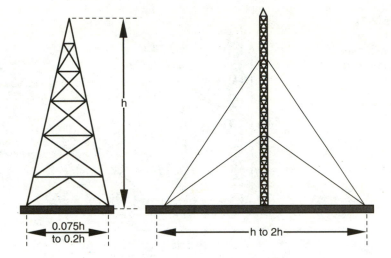

■■■ ■■■ ■■ ■■
Figure 2-12
Tower base
mounting

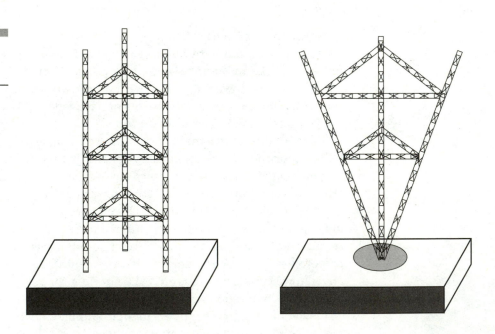

manufacturers. In many instances, a tower turnkey contract can be obtained that provides complete engineering, installation, grounding, lighting, and antenna installation services. Such a contract is often the best way to go for a cable system operator since most cable systems do not have a staff member trained in tower work.

Tower Grounding Methods Proper tower grounding is an important consideration in the control of lightning during a thunderstorm. After all, the tower is a big lightning rod itself. Some operators just ground the base of the tower to a system of ground rods. A base-grounding method is shown in Figure 2-13. In heavier lightning-prone areas, a sharp copper rod is mounted at the top of the tower and connects through heavy copper wire to the ground point at the base of the tower. Placing heavy copper strips in the concrete before the tower base is poured and then when the base has cured ties the four ends together. Placing the tower base on top usually provides a superior ground for the tower.

Testing the ground with an instrument known as a megger will indicate the effectiveness of the ground. If this is performed at the time of installation, subsequent measurements on a yearly basis will detect any deterioration of the tower ground system. If a tower requires lighting, a commercial electric service has to be connected all the way up the tower. Thus, a good and effective ground is necessary to protect the lighting system and any tower-mounted electronic devices such as pre-amps and converters.

Tower Lighting Tower lighting systems are usually available through the tower manufacturer. The flashing control system is usually mounted near the tower base on one leg of the tower and has a light-detecting cell

Figure 2-13
A typical tower grounding method

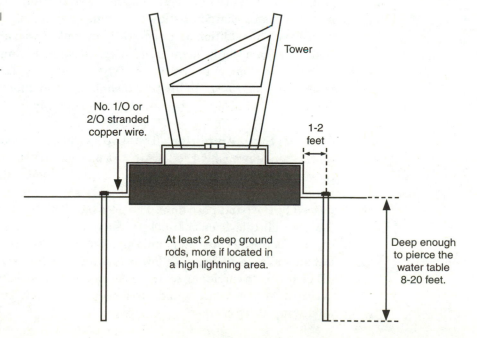

Tower

No. 1/O or 2/O stranded copper wire.

1-2 feet

At least 2 deep ground rods, more if located in a high lightning area.

Deep enough to pierce the water table 8-20 feet.

that turns the light system on at dusk. Since the tower lights use considerable electricity, it is prudent to just turn them on at dusk and off at sun-up. The flashing mechanism is also mounted in the control box. Early flashing systems were electromechanical where a motor turned a crank that opened and closed the beacon switch. Most present-day systems use solid-state flashing technology.

A periodic inspection of the tower system should be performed on a seasonal basis (four times a year). Such an inspection should include the guying system, as well as the anchors and wire fastenings and grounding. Any loose clasps should be tightened and any observable rust should be coated with a protectant spray. The lighting system can be easily checked by placing a heavy work glove over the daylight-sensing element, thus turning on the beacons and side lights. Burned-out bulbs should be promptly replaced. Also the tower grounding system should be inspected and any observable corrosion treated. A record of these inspections should be kept in a tower maintenance file.

Head-end Electronic Equipment

Signal-processing systems for received off-air television signals have changed drastically and for the better throughout the years. Early systems simply used single-channel amplifiers operating with vacuum tubes as the gain element. With the development of transistors that exhibited high-frequency gain characteristics, the single-channel strip amplifiers greatly improved. The addition of automatic gain control was also a welcome addition. A major improvement in off-air signal-processing methods occurred with the development of the heterodyne signal processor. Signal processors today use *integrated circuit* (IC) technology and therefore are physically smaller, consume less power, and operate superbly.

Channel Strip Amplifier The strip single-channel amplifier with automatic gain is probably the lowest cost method to use for off-air signal processing. Essentially, this method couples a gain block with channel band-pass filters. At this point, the only signal amplified is the channel passed by the band-pass filter. The pass band is usually flat throughout the 6-MHz-wide television channel. The signal level is adjusted by a front panel control. Both the video and audio signals are passed through the amplifier, but an audio 4.5-MHz trap allows the sound carrier to be set to the usual 15 dB below video carrier level. This is necessary for consecutive channel operation and prevents the sound carrier level from interfering with the upper adjacent picture carrier.

Figure 2-14
Strip amplifier

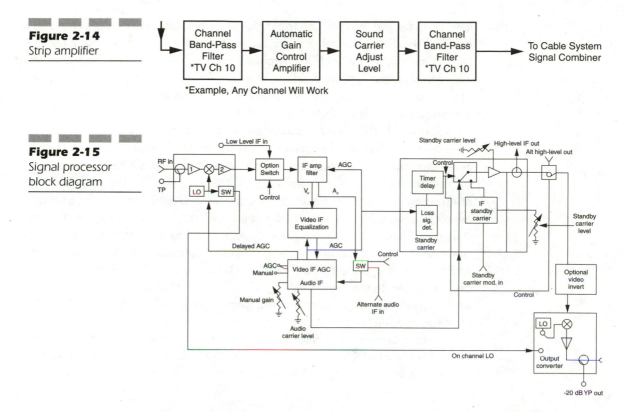

Figure 2-15
Signal processor
block diagram

Strip amplifiers often appeared in early systems and were often replaced later with the heterodyne signal processor. The strip amplifier method is shown in Figure 2-14.

The Heterodyne Signal Processor   Most systems use heterodyne signal processors for off-air television broadcast services. At one time, some cable systems opted for the Demod/Remod scheme, which required the received television station to be demodulated to base-band and remodulated back to either the same or a different cable channel. Such a system allowed local messages and or advertisements to be inserted at the video signal level. Also an *emergency alert system* (EAS) could be inserted in all the television channels. Voice or audio signal override could be easily introduced. Many manufacturers of present-day heterodyne signal processors offer many of the same features as the Demod/Remod method, only signal switching is done at IF. A block diagram of a heterodyne signal processor is shown in Figure 2-15.

This type of processor has many desirable features that enable many options for the operator. Probably the most desirable feature is the stand-by carrier. If, for any reason, the input signal is lost, either because the station went off the air or a problem occurred with the receiving antenna, a substitute carrier at the correct level would be introduced into the cable system. This replacement signal would continue until the off-air signal was returned to the input of the processor and this standby carrier could be modulated by a replacement video signal. Such a signal could be a simple "please stand by" message. If the channel happened to be used as a pilot carrier on the system, a properly level-adjusted stand-by carrier is absolutely required in order to maintain the cable system automatic level and slope system.

Another feature that is often used is alternate IF switching, which allows another signal source to be introduced at the IF frequency. This feature was developed for cable systems that were required to drop an off-air channel that contained the same program as one of the must-carry broadcast stations. Such a condition is referred to as non-duplication or simply non-dupe switching. Cable systems that use the off-air method of receiving broadcast programs use heterodyne signal processors. Several manufacturers make some superb signal-processing equipment that also features the capability of phase-locking to a comb generator. Such phase-locking is necessary for either an *incrementally related carrier* (IRC) system or a *harmonically related carrier* (HRC) controlled systems. A block diagram of the processor phase lock loop is shown in Figure 2-16.

Figure 2-16
Signal processor
phase-lock circuit

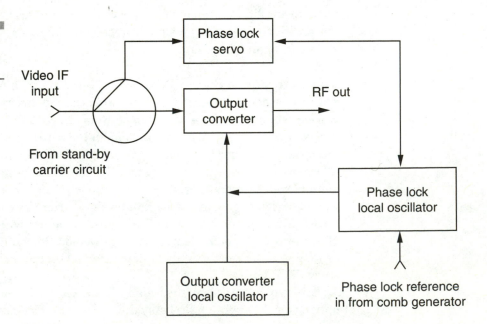

The Television Modulator   A television modulator, as used by the CATV operator, is in essence a miniature television broadcast transmitter. The signal-processing stages have to perform with the same specifications as a broadcast television station. The block diagram for a present-day television modulator is shown in Figure 2-17. Elementary television modulators are found in such devices as VCRs where the recorded video/audio signal is played and sent to a modulation circuit that transforms the video/audio signal to a standard TV channel, typically channel 3 or 4. The modulator for cable television use is vastly different from this type of device.

Referring to Figure 2-17, notice that there are two video inputs and two audio inputs. One is the main channel and the other is the alternate channel, which is controlled by a selector switch. The switch can also have the option to switch to the alternate source upon loss of the main input video signal. The alternate source can be a simple "please stand by" or some other appropriate message. Just about any signal can either be introduced or outputted for a variety of applications. Options provide for full video/audio metering and system phase-locking capability.

Manufacturers of head-end equipment offer cable systems a variety of television modulators from extremely compact types to larger ones containing a large selection of options. Some modulators have a selectable or tunable output and can be considered as a sort of universal device. Often these modulators do not have a selectable band-pass output filter and can

Figure 2-17
Television modulator

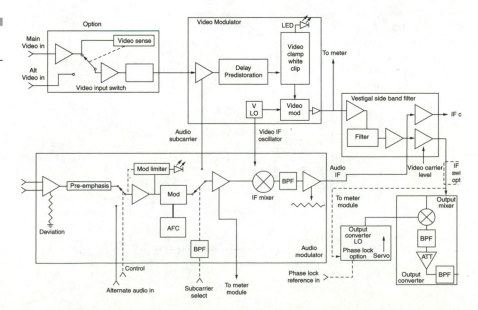

add broad-band noise into the channel-combining network. A more prudent approach is to use normal single-channel output converters with band-pass filtering for all channels requiring a modulator. The tunable type modulators can be used as spares. In general, the use of modulators is for satellite-originated programming and locally generated program channels.

Signal Combining and Filtering   The next process in assembling the television channel line-up is the combining of all the program channels into one terminal used to drive the coaxial cable network. Care must be taken at this point to not inject any added noise or unwanted signals into the system. In some instances, through either local ordinances or for economic reasons, the cable television system may use an active AM or FM tower for its off-air antenna system. Thus, the problems of preventing the transmitted station from entering the cable system can be extremely difficult. An example of such a problem is shown in Figure 2-18, but the solution may not be as simple as that shown. Situations such as this should be avoided if at all possible. A spectrum analyzer measurement of the whole cable spectrum at the combining point will show all the video and audio carrier levels as well as any extraneous signals.

The combining network can be made from a tree of 75-ohm two-port splitters that is used as a combiner or as a series of directional couplers. The tree method contains the least loss and therefore the signals to be combined

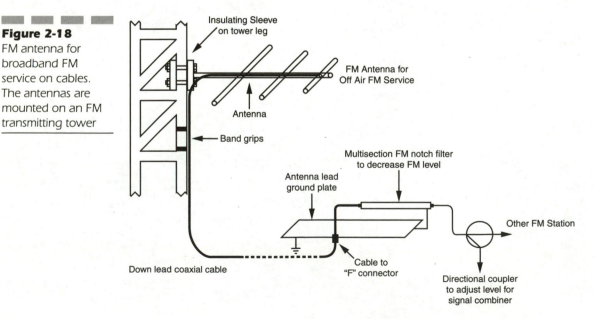

Figure 2-18
FM antenna for broadband FM service on cables. The antennas are mounted on an FM transmitting tower

do not have to run at such high levels. Manufacturers of head-end equipment usually have directional couplers for combining networks in their product lines. This is the more usual type of network used. Ideally, each channel should have an output band-pass filter. This prevents any extraneous broad-band noise from entering the cable system. The two mentioned types of combining methods are shown in Figure 2-19.

Head-end Powering and Monitoring

Cable television head-ends usually are operated by commercial power brought to the site by aerial or underground means. In the case where there is a commercial power failure, some means of backup power is recommended. Current cable systems should attempt to operate with the same reliability figures as the telephone system. This is a must-do if cable operators are going to compete in the business of providing voice, data, and video communications. Since signal reliability is an issue, head-end signal monitoring also should be considered.

Stand-by Power   Stand-by powering of head-end equipment falls into two basic categories: an engine-generator set and a battery back-up set. The

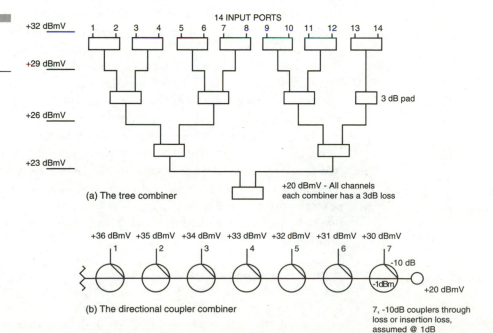

Figure 2-19
Signal combining
network types

(a) The tree combiner

+20 dBmV - All channels
each combiner has a 3dB loss

(b) The directional coupler combiner

7, -10dB couplers through
loss or insertion loss,
assumed @ 1dB

battery back-up set can be as exotic as an *uninterruptible power supply* (UPS). Since most head-ends have some kind of alarm system, such as for unauthorized intrusion, fire and/or smoke, these could also alert the loss of commercial power. Service personnel could then go to the head-end site to monitor the situation. The most common type of head-end emergency power is the engine generator set, simply because it can usually provide 100 percent of the usual power load that includes air conditioning and tower lights in addition to the normal electronic load. The engines are usually reciprocating multicylinder engines that operate on a variety of fuels. Large stand-by plants often use diesel fuel, while some use gasoline, propane, or piped-in natural gas as a fuel.

Several manufacturers of engine-driven stand-by power plants offer equipment in a broad range of powers. Battery-powered back-up systems usually provide enough power covering anywhere from a one- to two-hour off period. For engine-driven emergency power systems, an automatic program change-over switch is the method of choice. This device starts the engine, tests the voltage, and transfers the power. Upon return of commercial power, as sensed by the switch, the power transfer is made back to commercial power and the engine is shut down.

An engine-run hour meter records the amount of time the engine was in operation. This data should be recorded in the plant maintenance file. Some emergency electric plant controllers have a weekly or monthly cycle period where the engine starts and runs for a predetermined period and then shuts down. This so-called exercise cycle may or may not actually transfer the electrical load.

Battery back-up power systems usually do not have an exercise cycle. The condition of the battery pack should be checked visually as well as electrically, and the condition of the charge should be recorded in the system maintenance files. Most uninterruptible power back-up systems either remain online, providing power while simultaneously being charged, or are switched online at commercial power outages. At the return of commercial power service, these systems transfer off-line and resume charging. As yet, most manufacturers of head-end electronic devices do not offer back-up power options on a device-by-device basis. Since many computer systems use such back-up power supplies, it is most likely this method will be used in hubs and head-ends.

Signal Monitoring Manual or Automatic   Signal reliability is now of extreme importance in many modern systems, so some means of signal monitoring and testing should be used. The testing/monitoring program can be done manually by test technicians and monitoring data is recorded in a site logbook. The advantage of such a program is that if a problem is observed it can be corrected at once. However, this method, being labor-intensive, is the most expensive. An automatic method involves the installation of programmed instruments to make periodic measurements in a predetermined sequence. The results can be converted to digital data and recorded on a floppy disk. This system can also be interrupted manually at the control computer terminal for a visual test of the system's operating parameters.

If one or more head-end sites are monitored in this fashion, a central office–monitoring computer can be operated via a data communication channel through the cable system. Such an interconnected system is shown in Figure 2-20. Each site can in turn be interrogated for its measurement data by the control computer through the usual data-type modems. Data found to be faulty by the monitoring computer can signal an alarm so technical personnel can respond and correct the problem. Carefully selecting the alarm threshold can facilitate heading off problems before an actual signal outage occurs.

Grounding and Bonding　An important concern for the head-end electronic equipment is that grounding and bonding be done properly. For most head-end layouts, electronic devices such as signal processors, modulators, satellite receivers, and ad-insertion equipment are mounted in metal racks. In some instances, open-standard, 19-inch telephone-type racks are used, and for more upscale head-ends, the equipment is mounted in enclosed rack cabinets. The metal racks should be connected together with either heavy

Figure 2-20

Status monitoring of end levels and/or power supply status

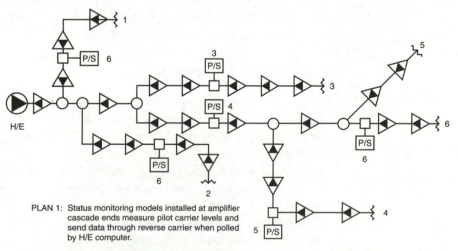

PLAN 1: Status monitoring models installed at amplifier cascade ends measure pilot carrier levels and send data through reverse carrier when polled by H/E computer.

PLAN 2: Test status of each power supply when polled by H/E computer.

PLAN 1 and PLAN 2 could both be used.

(1/0 or 2/0) copper wire, bus bars, or braids and then connected to a grounded electrode. This ground point should be the same electrode as the power system ground. Now no fault currents can set up through several paths to ground. The ground method should provide less than 25 ohms to the ground, as tested with a megger or milliohmmeter.

Some head-ends operating near television or FM/AM radio-transmitting facilities have to use RF chokes and bypass capacitors at the power service panel to prevent RF ingress of unwanted signals entering the cable system. Cables entering the electronic equipment building from the antenna, often referred to as downloads, have to be connected to a heavy aluminum plate grounded at the head-end common ground point. The heavy, usually half-inch solid aluminum sheath system cable used as a download is connectorized to this panel through half-inch "F" connectors. Now simple RG-6 cable connects to the equipment in the racks.

Since most head-ends use towers for the off-air receiving antennas, lightning can again be a problem. Proper grounding can help prevent the direct effects of a lightning strike, but large spikes of voltage can be induced into the power lines, causing power surges of significant magnitude. Such power surges can cause blown fuses or equipment failure of a significant nature. A good idea is to have a surge suppression system in place. Simple surge suppressors can be used at each device plug, but a larger unit mounted in or at the power distribution panel saves a lot of clutter and is quite effective. Several manufacturers make panel-mounted surge suppressors, offering various features, shapes, and sizes. Basically, the surge suppressor has to pass the load, removing any voltage spikes and delivering smooth power free of electrical transients to the equipment. By now, it should be evident that there are a large number of considerations when planning a head-end.

Satellite Systems

Some of the people who started cable systems back in the 1950s and 1960s recall the days when there were no satellite systems operating. Cable television systems were simply a means used to deliver greatly improved television broadcast signals to subscribers, a true community antenna television system. Many early systems had unused bandwidth or channel space. When satellite systems were developed to the point where more programming became available to the cable operator, more bandwidth was needed to carry the added services. The axiom that said, "If you have more to sell, you sell more," became an important rule for cable operators. Communication satellites are very important to the telecommunication industry and affect many users around the world.

Geostationary Satellites Of the many satellite systems in place covering the Western hemisphere, the geostationary ones are the important ones to the cable television industry. Basically, a geostationary satellite is in a circular, equatorial orbit. Figure 2-21 shows the geostationary satellite concept. Notice that the usual orbit direction is the same as the earth's rotation and the height above the earth surface is about 36,000 kilometers (21,600 statute miles). One such satellite can cover better than two-thirds of the earth's surface. Satellites that orbit the earth have to conform to the velocity equation given in Figure 2-21. Satellites that are in a geostationary orbit have the same angular velocity as that of the earth. Therefore, the circumference of the orbit divided into the orbit speed as calculated by the equation will give the period of the orbit as 24 hours. This is the same as the earth's period of rotation.

When planning a satellite receiving system for a cable system, a frequency coordination study should be made as soon as the desired satellite channels have been determined. This study provides information on the amount of expected or possible interference from any other surrounding microwave communication facilities. Microwave-common carriers, such as any of the local telephone systems, use the same frequency bands as some of the satellite systems. This study is a service offered by several companies that has amassed a database of frequencies and their propagation paths. If this study indicates possible interference from any local sources, an onsite measurement may be necessary to determine if the interference is of a sufficient level to cause a problem. A possible cure is to erect a conductive shield near the receiver antenna or to construct an artificial hill in the direction of interference. Some manufacturers make a shield ring around

Figure 2-21

Geostationary satellite orbit

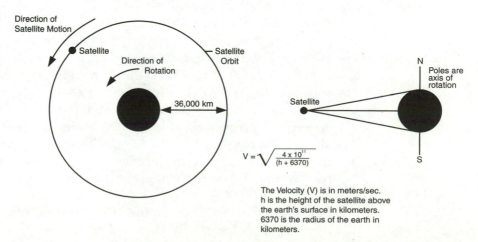

$$V = \sqrt{\frac{4 \times 10^{11}}{(h + 6370)}}$$

The Velocity (V) is in meters/sec.
h is the height of the satellite above the earth's surface in kilometers.
6370 is the radius of the earth in kilometers.

the parabolic antenna, which reduces the effective side lobe pattern of the antenna. If no cures are feasible, the receiving site may have to be relocated to a place that has no interference problems. Now it may be necessary to super-trunk the signals to the hub head-end site or main signal distribution location. All of this investigative work should be satisfactorily completed before any construction begins.

Siting the Earth Station   One of the important tasks is to make sure the parabolic receiving antenna can get a clear path to the satellite. No trees or buildings can obstruct the path. In many instances, some trees have to be removed so the path from antenna to satellite is clear. In order to do this, the azimuth and elevation look angles have to be determined.

Several pieces of information have to be assembled in order to establish the look angles for the antenna. The satellite longitude has to be known, but notice there is no latitude given because zero latitude is the equator. Next, the receiving site latitude and longitude have to be determined to an accuracy of the nearest hundredth of a degree. A U.S. coast and geodetic survey map of the 7.5-minute type is adequate to determine the site latitude/longitude coordinates. Often sporting goods stores have these maps for sale. A local civil engineering/surveying office will be able to determine the site coordinates. A GPS hand-held satellite global positioning receiver operating in the differential mode can provide coordinate information to a sufficient accuracy. Such instruments are available in sporting goods stores for a few hundred dollars. In most cases, the prospective site coordinates are needed for the frequency coordination study. The calculation for the azimuth angle in degrees from true north and the elevation angle in degrees from level ground can be calculated. The method and formulas with a worked-out example are shown in Example 2-1.

Once the elevation angle is found, the slant range or simply the distance from the receiving antenna site to the satellite can be determined, as shown in Example 2-2. This distance is used to calculate the signal path loss.

For various satellites, the satellite communication operators have a map with the projected *effective isotropic radiated power* (EIRP) contour lines appearing on it. For receiving sites found on the map, the EIRP can be determined. An example of such a map is shown in Figure 2-22. The calculation for the expected carrier-to-noise level at the receiver pre-amplifier input can now be made and an example is shown in Example 2-3. If a better C/N ratio is required, a larger antenna will be required.

Satellite Receiving Antennas   Once the calculations are made, the receiving antenna specifications will be known. Now the decision can be

■■ ■■ ■■ ■■
Example 2-1
Satellite pointing
angles

The formulas are as follows:
Antenna azimuth angle (degrees) =

$$180 + \tan^{-1} \frac{(\tan D)}{(\sin X)} \text{ (from true north)}$$

Antenna elevation angle (degrees) =

$$\left(\frac{\{\cos D \times \cos X\} - 0.15126}{\sqrt{\{\sin D\}^2 + \{\cos D \times \sin X\}^2}} \right)$$

where X = earth station site latitude
      Y = earth station site longitude
      Z = satelite longitude
      D = Z − Y

Example

Earth station site latitude 41° 38' 11.0" N
Earth station site longitude 70° 27' 42.0" W
Satelite galaxy I longitude 93.5° W
Solution: Convert latitude and longitude of Earth station site to decimal degrees.

Earth station latitude		Earth station longitude	
41° =	41.0000	70° =	72.0000
38' = 38/60 =	0.6333	27' = 27/60 =	0.4500
11.0" = 11/3600 =	0.0031	42" = 42/3600 =	0.0117
	41.6364 = 41.64		70.4617 = 70.46

where X = 41.64°
      Y = 70.46°
      Z = 93.5°
      D = 93.5 − 70.46 = 23.04°

Azimuth angle (degrees) $= 180 + \tan^{-1}\left(\dfrac{\tan D}{\sin X} \right)$

$$= 180 + \tan^{-1}\left(\frac{\tan 23.04°}{\sin 41.64°} \right) = 180 + \tan^{-1}\left(\frac{0.425}{0.66} \right)$$

$$= 180 + \tan^{-1}(0.644)$$

$$180 + 32.8° = 212° \text{ from true north}$$

Elevation angle (degrees) $= \tan^{-1}\left(\dfrac{\{\cos D \times \cos X\} - 0.15126}{\sqrt{\{\sin D\}^2 + \{\cos D \times \sin X\}^2}} \right)$

$$= \tan^{-1}\left(\frac{\{\cos 23.04 \times \cos 41.64\} - 0.15126}{\sqrt{\{\sin 23.04\}^2 + \{\cos D \times \sin X\}^2}} \right)$$

$$= \tan^{-1}\left(\frac{\{0.920 \times 0.747\} - 0.15126}{\sqrt{\{0.391\}^2 + \{0.920 \times 0.664\}^2}} \right)$$

$$= \tan^{-1}\left(\frac{0.6872 - 0.15126}{\sqrt{0.153 + 0.373}} \right)$$

$$= \tan^{-1}\left(\frac{0.5359}{0.723} \right) = \tan^{-1}(0.7389) = 36.5°$$

Example 2-2

Calculation of the distance from the receiving antenna site to the satellite

The distance from a point on Earth to a Satelite can be calculated using the following formula:

$$d = \sqrt{[(r + h)^2 - (r(\cos\Theta))^2]} - r(\sin\Theta)$$

d = distance to the satellite in km
r = radius of the earth = 6370 km
h = height above satellite above equator = 35780 km
Θ = elevation angle in degrees

Example: Θ = 36.5°

$$d = \sqrt{(42.15 \times 10^3)^2 - ((6.37 \times 10^3)(\cos 36.5°))^2} - (6.37 \times 10^3)(\sin 36.5°)$$
$$= \sqrt{1806 \times 10^6 - (5.12 \times 10^3)^2} - 3.79 \times 10^3$$
$$= \sqrt{1806 \times 10^6 - 26.2 \times 10^6} - 3.79 \times 10^3$$
$$= \sqrt{1780 \times 10^6} - 3.79 \times 10^3$$
$$= 42.2 \times 10^3 - 3.79 \times 10^3 = 38.4 \times 10^3 \text{ km}$$
$$= 38400 \text{ km}$$

Figure 2-22

Calculation of the distance from the receiving antenna site to the satellite

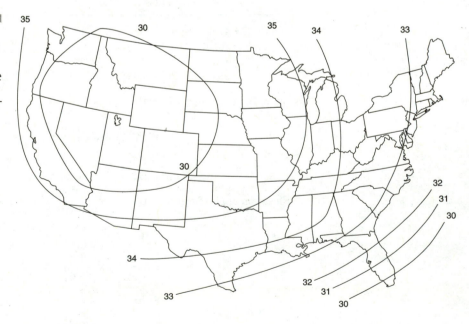

Example 2-3
Calculation of
path loss

To calculate the path between a satellite and a receiving site on earth may be calculated using the following procedure:

$$L_{path}(dB) = 32.44 + 20\log d_{(km)} + 20\log f_{(mHz)}$$

Example: for a frequency of 4 GHz (4000 MHz) and where d = 38400 km.

$$L_{path}(dB) = 32.44 + 20\log 38400 + 20\log 4000$$

$$= 32.44 + 20\log 3.84 + 80 \text{ dB} + 20\log 4 + 60 \text{ dB}$$

$$= 32.44 + 11.69 + 80 \text{ dB} + 12 + 60$$

Path loss = 196 dB

made as to the antenna size, the low noise block-down converter, and the types of feed-mount and lead-in cable. Several types of antennas can be fitted with feed mounts covering several satellites depending on their spacing in the satellite arc. For certain applications, five- and six-meter antennas can be fitted to cover three different fairly closely spaced satellites. Since adjacent satellite channels are transmitted alternately polarized (odd-numbered channels horizontally polarized with even-numbered channels vertically polarized), the frequency difference between adjacent channels and this polarization difference prevents any adjacent channel interference. Table 2-1 lists some commonly used C-Band channel information.

Satellites with polarization listed in Column A are GE Satcom C3, C4, and Hughes Galaxy IX.

Satellites with polarization listed in Column B are Hughes Galaxy IV, V, VI, VII, I-R, G.E. Satcom C1, and G.E. American GE-1.

Satellite receiving antennas have been fitted with the required vertically and horizontally polarized feeds with a low noise block-down converter connected to each one. The converter performs two basic functions. The first is to amplify the weak microwave signal and the second is to convert the

Table 2-1

C-Band Satellite
Channel
Information

Sat. Ch.	Down Link MHz	Down Link Polarization	
		A	B
1	3720	V	H
2	3740	H	V
3	3760	V	H
4	3780	H	V
5	3800	V	H
6	3820	H	V
7	3840	V	H
8	3860	H	V
9	3880	V	H
10	3900	H	V
11	3920	V	H
12	3940	H	V
13	3960	V	H
14	3980	H	V
15	4000	V	H
16	4020	H	V
17	4040	V	H
18	4060	H	V
19	4080	V	H
20	4100	H	V
21	4120	V	H
22	4140	H	V
23	4160	V	H
24	4180	H	V

signals to a lower frequency. Converting to a lower frequency results in lower cable loss and therefore a less attenuated signal at the receiver. This method is shown in Figure 2-23 and is the most commonly used method.

Figure 2-23

Satellite antenna
block converter
configuration

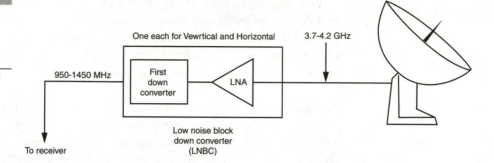

One each for Vewrtical and Horizontal

3.7-4.2 GHz

950-1450 MHz

First
down
converter

LNA

To receiver

Low noise block
down converter
(LNBC)

Table 2-2

C-Band to L-Band
Conversion

Transponder Number	C- Band Freq. MHz	L-Band Freq. MHz	Transponder Number	C- Band Freq. MHz	L-Band Freq. MHz
1	3720	1430	13	3960	1190
2	3740	1410	14	3980	1170
3	3760	1390	15	4000	1150
4	3780	1370	16	4020	1130
5	3800	1350	17	4040	1110
6	3820	1330	18	4060	1090
7	3840	1310	19	4080	1070
8	3860	1290	20	4100	1050
9	3880	1270	21	4120	1030
10	3900	1250	22	4140	1010
11	3920	1230	23	4160	990
12	3940	1210	24	4180	970

Satellite Receivers The receiver takes the converted C-band signals and produces a base-band descrambled signal. Many manufacturers provide what is known in the industry as an IRD (integrated receiver descrambler) satellite receiver. The receiver descrambler circuits are integrated into one package, thus conserving rack space in often crowded electronic equipment bays in larger head-ends. Table 2-2 shows a listing of C-band channel frequencies to their down-converted values. Notice that the low-frequency

C-band signals are converted to high-band frequency signals. This is due to the heterodyne frequency conversion process.

The L-band receivers on the market today have the C-band channel number indicated on the front panel, making the frequency values superfluous. An L-band receiver is shown in Figure 2-24.

Pay (Premium) TV Systems   Much of the programming on satellite channels is scrambled and/or encrypted using several methods. The *General Instrument Corporation* (GI) developed the earlier type called Videocipher II, which was upgraded to Videocipher II+. Later methods that were developed and used by GI are DCI, DCII, DCII MCPC, and DCII SCPC. Scientific Atlanta Corporation has a series that is available to program suppliers called Pwrvu, Pwrvu MCPC, and Pwrvu SCPC. The purpose of using any of these methods of encoding is to prevent the unauthorized use of the programming material. The key for decoding is contained in the authorized equipment in the cable system head-end. The cable operator can then rescramble in the format compatible with the system's subscriber converter box.

A general observation might raise the question, if the signal is already scrambled, why descramble and then rescramble? The obvious answer depends on who is doing the scrambling or protecting the program service. Since passing the signal flow through more circuits adds noise or distortion,

Figure 2-24
Base band
output signals

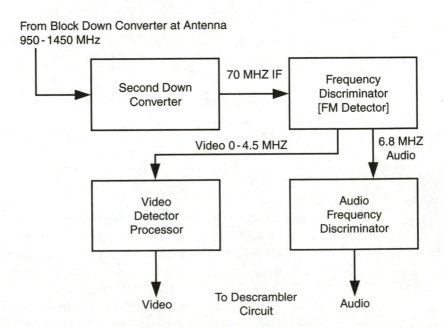

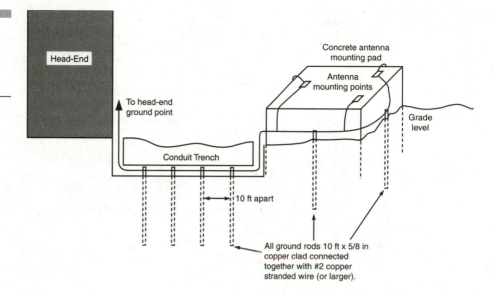

Figure 2-25
Satellite antenna
mounting platform
and grounding
system

possibly a better arrangement would be to pass the scrambled signal through the cable system to the subscriber's set-top box, where the signal is programmed to decode or unscramble. Doing it this way, the program supplier knows the cable company that is providing the service to the subscriber.

Grounding the earth-station antennas is also important in order to protect the sensitive LNBC equipment from the effects of lightning. Such a grounding plan is shown in Figure 2-25.

Satellite Signal Processing   At the master head-end site, the satellite-received signals are either sent clear as part of a basic program package or are rescrambled as a premium or pay channel. All satellite channels are modulated on a cable system channel and are combined with the off-air broadcast stations plus any of the locally generated channels. In general, locally generated programs are the local television studio-originated signals, a character generator with the listing of the programs (program guide), and a character generator with lost-and-found column news announcements.

A block diagram showing the device interconnections for the processing of satellite program signals is shown in Figure 2-26. As suspected, many options and choices are available within the methods of interconnecting the equipment required to assemble the program line-up. Whatever the choice, the signal quality must always be preserved simply because it can never be better at the head-end. Once the signal enters the cable system, it is naturally degraded and much now depends on the cable system design.

Figure 2-26

13 channels from one satellite transponder, 6 pay channels re-scrambled and 5 clear chanels

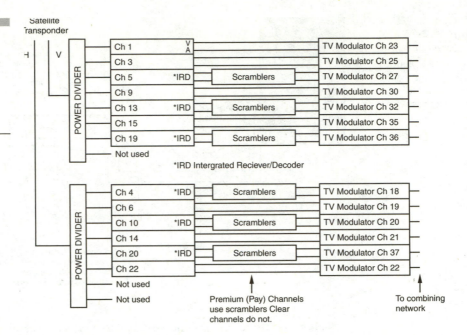

The Cable Distribution Plant

The coaxial cable system design consists of the trunk and distribution of coaxial cable sections of cascaded amplifiers. As mentioned before, simpler elementary systems of earlier designs consisted of single cable-connected amplifier cascades. No differences were made between trunk and feeder cables. It was soon determined that the trunk cable cascade should be the signal transportation medium and should just supply the signal to the feeder cable system containing subscriber taps and distribution type (line extender) amplifiers. This method extended the signal the greatest physical distance from the head-end and became known as the trunk-feeder system, or rather simply the trunk/branch method.

The Trunk-feeder System

The feeder-trunk method became the most accepted type of cable system design, known as the bread and butter of system architecture. System design and layout was often done manually on a drafting table, overlaying the design on a series of street maps. Often maps can be obtained from several possible sources, such as:

1. The local telephone or power company

2. A local land surveyor/civil engineer

3. The local town or city department of public utilities

4. U.S. Coast and Geological Survey maps

Many utility companies also offer their maps on either floppy disk or CD-ROM. Once maps are obtained, then strand-mapping can commence.

Since maps may not contain any late changes, a measurement using a measuring wheel or a surveyor's transit and stadia rod to measure the distance between utility poles may be necessary. If utility poles are not going to be used, then a wheel measurement along the buried plant route should be made. For an aerial plant on utility poles, a pole attachment agreement with the pole owners should be initiated and in the works. Once strand mapping is complete, make-ready work can commence. Since the measuring wheel is a low cost and a simple means of establishing the cable system length, cable operators should keep several handy for measuring added plants as well as subscriber drop lengths. Such a wheel is shown in Figure 2-27.

When the measurements are completed and the distances are entered on the map, then the poles and the route of the cable plant are known. At this

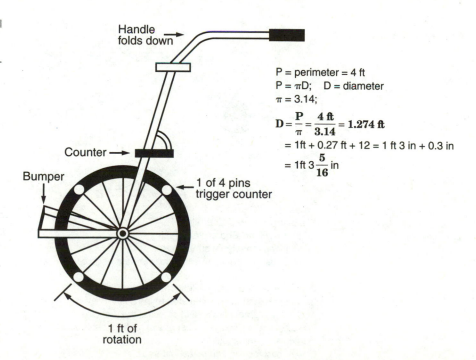

Figure 2-27
Measuring wheel

Handle folds down

Counter

Bumper

1 of 4 pins trigger counter

1 ft of rotation

P = perimeter = 4 ft
$P = \pi D$; D = diameter
$\pi = 3.14$;

$$D = \frac{P}{\pi} = \frac{4\ \mathrm{ft}}{3.14} = 1.274\ \mathrm{ft}$$

$$= 1\mathrm{ft} + 0.27\ \mathrm{ft} + 12 = 1\ \mathrm{ft}\ 3\ \mathrm{in} + 0.3\ \mathrm{in}$$

$$= 1\mathrm{ft}\ 3\frac{5}{16}\ \mathrm{in}$$

time, two projects can be completed at the same time. The first is the make-ready survey with the pole owners and is often a three-party survey between the cable operator representative and representatives of the telephone and electric companies. Each pole that is to be used for the cable plant is examined to determine if there is proper space for the cable system. If any of the telephone or electric wires have to be moved to make room for the cable system, the cost is charged to the cable system. Most cable systems are well aware and are experienced in the use of leased space on the local utility poles. For underground or buried cable plants, only permission from the municipality is needed to perform the work. Responsibility for any damage to an existing buried plant such as telephone, power, water, sewage, and gas falls on the cable operator. It is usually the responsibility of each plant owner to mark-out their respective plant on the request of any of the other companies.

The Trunk System　Once the system routing has been established, the trunk layout can be designed. This can be projected by the engineering department based on known areas of system expansion. An example of a trunk layout is shown in Figure 2-28. From this point, a system design can

Figure 2-28
Trunk amplifier routing needed to cover the service area

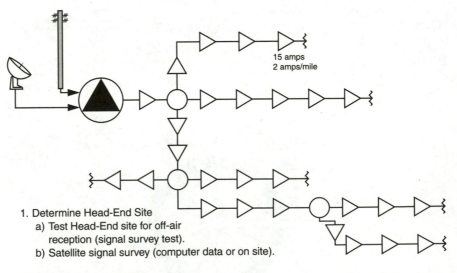

15 amps
2 amps/mile

1. Determine Head-End Site
   a) Test Head-End site for off-air reception (signal survey test).
   b) Satellite signal survey (computer data or on site).

2. Define service area - miles/roads.
3. Draw skeletal trunk route, determine miles.
4. Calculate amplifiers/mile of selected trunk cable.
5. Go back and make any changes in routing, cable size to cover plant layout.
6. Now cost factors can be considered.

either be done manually using a pocket calculator or a more sophisticated computer. Several versions of software are available for personal computers and some are more involved and complicated than others. Computer-aided design and drafting programs have been around for some time. Most current ones can produce simultaneous trunk/feeder designs complete with tap values, port signal levels, amplifier input/output levels, and noise and distortion values, as well as a complete bill of materials. Since such software, computer, and plotting equipment is expensive, it is more cost-effective to have such work contracted out. Currently, several such design contractors exist throughout the country and most provide cost-effective designs in a timely manner.

The Bridging to Feeder System   The feeder or distribution system contains the subscriber taps, which are the part of the system that produces revenue. Thus, it is extremely important that it is also properly designed and constructed. Essentially, the feeder system is bridged off the trunk system through a bridging amplifier module installed in a trunk amplifier housing. From this bridging amplifier, the distribution cable delivers a signal to a series of subscriber taps. When this signal level decreases sufficiently, a line extender or distribution amplifier is required to increase the signal level for another series of taps. Most wide-band 750-MHz systems limit the number of distribution amplifier cascades to two amplifiers. Figure 2-29 shows a section of a distribution system, displaying what happens to signal levels along the distribution cable plant.

Notice that at the bridger output the signal has a 9-dB slope and, due to taps through loss and normal cable loss, the signal level at the last tap port nearly has a 9-dB tilt (reverse of slope). The higher the upper frequency limit becomes, the more severe the tilt becomes. Thus, the insertion of an inline equalizer before the last tap will provide signal correction. If the system is to be extended beyond the last tap, however, a line extender amplifier with a proper equalizer will be required. Normally, most systems limit the number of line extender amplifiers to two in a cascade. This results in three cable sections of five or six four-port taps, servicing 65 or 70 homes.

As the upper-band limit is expanded from 750 MHz on up to 1 GHz, the spacing between amplifiers for both the trunk and feeder systems shrinks. Now more amplifiers per mile are required as well as more power supplies to feed them. The constant fear is that the build-up of noise and amplifier distortion products would limit the system length or reach. In some instances, it becomes possible to change system routing to control noise and distortion problems, but nothing can be done about the high power consumption of a wide-band coaxial cable system plant.

Figure 2-29

Example of a
distribution section
of a cable plant

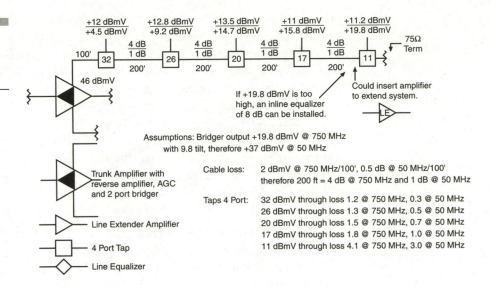

| | +12 dBmV | +12.8 dBmV | +13.5 dBmV | +11 dBmV | +11.2 dBmV | |
| | +4.5 dBmV | +9.2 dBmV | +14.7 dBmV | +15.8 dBmV | +19.8 dBmV | 75Ω Term |

100'   [32]   4 dB / 1 dB   [26]   4 dB / 1 dB   [20]   4 dB / 1 dB   [17]   4 dB / 1 dB   [11]

46 dBmV   200'   200'   200'   200'

If +19.8 dBmV is too
high, an inline equalizer
of 8 dB can be installed.

Could insert amplifier
to extend system.

LE

Assumptions: Bridger output +19.8 dBmV @ 750 MHz
with 9.8 tilt, therefore +37 dBmV @ 50 MHz

Trunk Amplifier with
reverse amplifier, AGC
and 2 port bridger

Cable loss: 2 dBmV @ 750 MHz/100', 0.5 dB @ 50 MHz/100'
therefore 200 ft = 4 dB @ 750 MHz and 1 dB @ 50 MHz

Taps 4 Port: 32 dBmV through loss 1.2 @ 750 MHz, 0.3 @ 50 MHz
26 dBmV through loss 1.3 @ 750 MHz, 0.5 @ 50 MHz
20 dBmV through loss 1.5 @ 750 MHz, 0.7 @ 50 MHz
17 dBmV through loss 1.8 @ 750 MHz, 1.0 @ 50 MHz
11 dBmV through loss 4.1 @ 750 MHz, 3.0 @ 50 MHz

— Line Extender Amplifier

— 4 Port Tap

— Line Equalizer

Cable System Splitting and Coupling Cost- and signal-effective branching of a cable television distribution system is very important. Maintaining proper signal level at the tap output port is of utmost importance to
subscriber service quality for 750 MHz and higher systems. The approxi-
mate loss of RG-6 90 percent braid cable is 5.5 to 6 dB/100 feet. Therefore,
a system should have a tap port level of at least +10 dBmV at the 750
MHz frequency. Proper drop cable shielding effectiveness is also necessary
to control signal leakage and ingress.

You will often have several choices for branching within a service area
design. If a manual design method is being used, the designer can ponder
the various choices, taking signal level and cost factors into consideration.
Signal splitters and directional couplers are the passive devices used for
system street branching. Figure 2-30 and 2-31 illustrate two methods serving the same residential street layout. One method employs a balanced output splitter and the other one uses directional couplers. Figure 2-30, using
two direction couplers, has two short branches with one four-port tap each
and one long branch with two taps before a line entender is required. Figure 2-31 on the other hand could use two taps in three directions with the
last tap a terminating tap before a line extender amplifier is required to
extend the system.

Computer-aided design and drafting programs keep a running tabula-
tion of signal levels. When this level approaches a predetermined level, then
an amplifier is needed at that point to continue the cable run. Where

Figure 2-30

Directional couplers used to branch to A, B, and C

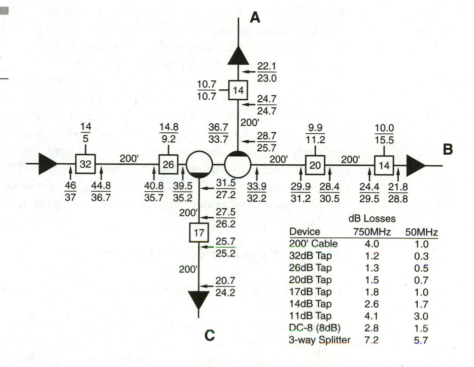

dB Losses		
Device	750MHz	50MHz
200' Cable	4.0	1.0
32dB Tap	1.2	0.3
26dB Tap	1.3	0.5
20dB Tap	1.5	0.7
17dB Tap	1.8	1.0
14dB Tap	2.6	1.7
11dB Tap	4.1	3.0
DC-8 (8dB)	2.8	1.5
3-way Splitter	7.2	5.7

Figure 2-31

Balanced 3-way splitter for branches A, B, and C

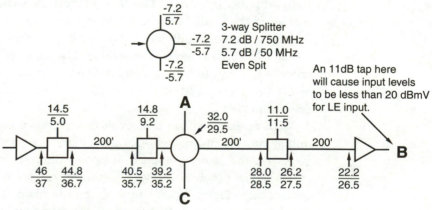

Note: All signal levels are +dBmV
         Top figure is @ 750 MHz
         Bottom figure is @ 50 MHz
         Legs A, B, and C are identical

branching is required, the designer operating the CAD system can make the branching choices. When an amplifier is inserted, the program then annotates the system map with such parameters as input-output level, pad value, equalizer value, and carrier-to-noise and carrier-to-distortion calculations. These parameters can be very useful when compared to actual system measurements at initial proof of performance time. Trunk system splitting and coupling is equally important in determining the system network topology. A system designer should have a good general idea where any areas of future expansion might be proposed. Designing in some additional trunking during an initial build can often pay off down the road when a new area is developed. This information can only be known by the people responsible for the system service area determination.

Cable System Electronics

Cable system electronics have come a long way from where it started, from vacuum tubes to transistors and now to *integrated circuits* (ICs). Today's cable amplifiers have more gain, higher output operating levels, improved noise and distortion figures, and less power consumption. Equally important is the high construction quality that gives the industry a superior, more reliable device.

The industry has also become more knowledgeable about the metallurgy involved in cable systems, resulting in corrosion-resistant and weather-tight amplifier housings. Modular construction of the amplifier sections has increased the flexibility and serviceability of the present-day cable amplifiers. Most cable amplifiers can be operated at more than one power supply voltage. Most systems operate at 60 volts a.c. and more progressive systems operate at 90 volts a.c. Jumpers in the power supply section can be set to select the proper system voltage. Also the stability of the amplifiers has been improved.

With proper thermal compensation circuits along with automatic gain and slope systems, present-day amplifiers can provide many years of service. In many cases, a properly designed and constructed cable system will cause so few problems, service and technical personnel do not get enough service experience. When things do break down, a lot of head scratching results trying to remember where to start and what to do.

Cascaded Amplifier Theory   The theory of cascaded amplifiers has been well known for many years. Radio communications people worked with short cascades to solve antenna pre-amplifier and strip amplifier signal distribution problems. Receiver designers used the theory to design multistage IF amplifiers. Telephone company engineers used cascaded repeater amplifiers in long-haul or long trunk lines where telephone voice channels were multiplexed on a coaxial cable. Double side-band AM techniques were used where one voice channel occupied the upper side-band and another voice channel occupied the lower side-band of a carrier. The carriers were in the 10 to 100 KHz range and groups of channels were formed according to CCITT international specifications. Telephone traffic was routed through coaxial cable–cascaded amplifier systems for much of the transcontinental runs. This technique was also used in microwave radio relay systems where the channel groups operating in the KHz bands were frequently converted to microwave carriers and were relayed by microwave hops. The microwave carriers could be converted to the lower frequency ranges used on the coaxial cable system. Many of the early cable television engineers and technicians found a lot of information about cascaded amplifier techniques in the technical papers published by the telephone companies.

Cascaded amplifier fundamentals appear in Appendix A. Most people familiar with the theory use the mathematical formulas shown in Figure 2-32 and often work with them, designing system additions and extensions using a scientific pocket calculator.

As the upper frequency limit of cable systems increased, and performance specifications as well, the number of amplifiers in cascades increased. This should be easily understood since the cable loss increased as the service bandwidth increased, so more amplifiers were needed per mile. Also, as better carrier-to-noise ratios and carrier-to-distortion figures were needed, the number of amplifiers allowed in a cascade decreased. Such problems

Figure 2-32

Noise and distortion formulas for a cascade of "n" amplifiers

Noise: $(C/N)_n = (C/N)_{n=1} - 10 \log n$

Composite second
order distortion (CSOD): $(C/CSOD)_n = (C/CSOD)_{n=1} - 10 \log n$

Composite
triple beat (CTB): $(C/CTB)_n = (C/CTB)_{n=1} - 20 \log n$

Cross modulation
distortion (XMOD): $(C/XMOD)_n = (C/XMOD)_{n=1} - 20 \log n$

n = number of amplifiers in a cascade

required many older cable systems to limit their bandwidth and the number of service channels. This problem is made clear by Example 2-4 where the upper frequency is 750 MHz with 110 channel loading. The sample calculations indicate that the amplifier cascade has to be limited to about five amplifiers if the specifications for distortion are to be adhered to. Clearly long cascades are prohibited for wide-band high-quality cable systems. Essentially replacing the coaxial cable trunk cascades with fiber optical systems enables cable operators to provide 110 channels at 750 to 1,000 MHz.

Trunk Amplifiers ASC/AGC   Trunk amplifiers then became available with a variety of characteristics such as push-pull, power-doubling, and feed-forward. Push-pull amplifiers provided the cable operators with an amplifier that had good second order distortion characteristics. Adding *automatic gain* (AGC) and *automatic slope control* (ASC) provided cable operators with the workhorse of the industry. Several manufacturers offered such push-pull AGC/ASC amplifiers. Power-doubling techniques, as shown in Figure 2-33, gave the industry higher output levels needed to extend system reach.

As bandwidth requirements increased and system length as well, the feed-forward amplifier was developed. This technique is shown in Figure 2-34 and it does for third-order distortion what push-pull did for second order distortion. Some cable operators extended cable cascades to nearly 45 amplifiers in a cascade for a 450-MHz 60-channel cable plant. Such systems operated at either an IRC or an HRC that allowed the third-order distortion levels to be relaxed by approximately six to nine dB.

Essentially, any distortion products were moved to the edge of the television screens, which was less objectionable. Cable system head-ends had to be phase-locked to a reference comb generator, causing the cable operator a lot of expense. Each modulator and signal processor had to have phase-lock circuitry added so they could be phase-locked to the reference generator. Often phase-locking at the head-end, as well as the use of feed-forward amplifiers, was necessary for cable operators to increase bandwidth and system reach.

Bridging and Distribution Amplifiers   Bridging and distribution amplifiers operate the feeder cable system containing the subscriber taps. It is imperative that the proper signal level at the tap output ports be maintained. Systems operating at 750 MHz with 110 channel loading will have difficulty feeding long subscriber drop lengths with adequate signal levels.

Example 2-4

750 MHz, 110
channel loading

For a single amplifier with a noise figure
of 10 dB and an input signal level of +15dBmV,
can be found using the following formula:

$$(C/N)_{n=1} = 59 \text{ dBv} + \text{signal input (dBmV)} - \text{noise figure}$$
$$= 59 + 15 - 10 = 64 \text{ dB}$$

To find 'n' for an end of cascade C/N of 49 dB.

$$(C/N)_n = (C/N)_{n=1} - 10 \log n$$

$$49 = 64 - 10 \log n$$
$$-15 = 10 \log n \quad \log n = 1.5$$
$$n = 32 \text{ amplifiers}$$

For a carrier to second order distortion specification
of 50 dB, the cascade number can be calculated by:

$$(C/SOD)_n = (C/SOD)_{n=1} - 10 \log n$$

For a single amplifier C/SOD = 63 dB

$$50 \text{ dB} = 63 \text{ dB} - 10 \log n$$
$$-10 \log n = -13$$
$$\log n = 1.3$$
$$n = \log^{-1} 1.3$$
$$n = 20$$

For a C/CTB and C/XMOD of 50 dB.

$$(C/CTB)_n = (C/CTB)_{n=1} - 20 \log n$$

For $(C/CTB)_{n=1} = 64$ dB

$$50 \text{ dB} = 64 \text{ dB} - 20 \log n$$
$$-14 = -20 \log n$$
$$\log n = 0.7$$
$$n = 5$$

For $(C/XMOD)_{n=1} = 65$ dB

$$50 \text{ dB} = 65 \text{ dB} - 20 \log n$$
$$-15 = -20 \log n$$
$$\log n = 0.25$$
$$n = 5$$

▮▮ ▮▮▮ ▮▮▮ ▮▮

Figure 2-33
Power doubling
amplifier

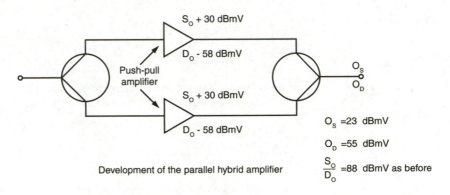

$$\frac{S_o}{D_o} = +30 - (58) = 88 \text{ dB} \qquad \frac{S_o}{D_o} = 88 \text{ dB}$$

Single Push-Pull-Type amplifier

Development of the parallel hybrid amplifier

$O_s = 23$ dBmV

$O_D = 55$ dBmV

$\dfrac{S_o}{D_o} = 88$ dBmV as before

S_o of parallel hybrid amplifier is 3 dB higher
with same $\dfrac{S_o}{D_o}$ ratio as a single amplifier.
A 3 dB improvement is eqivalent to doubling
the power.

▮▮ ▮▮▮ ▮▮▮ ▮▮

Figure 2-34
Feed forward
amplifier operational
example

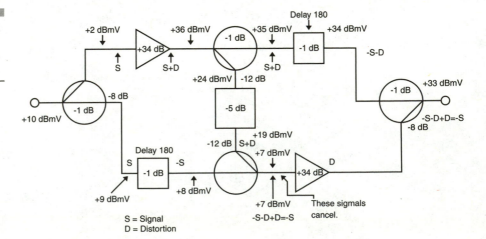

S = Signal
D = Distortion

Since the trunk amplifier's purpose is signal transportation, it is the job of bridging and distribution amplifiers to distribute the signal. The bridging amplifier module is installed in a trunk station housing and has the map symbol configuration, as shown in Figure 2-35.

The distribution cable system consisting of subscriber taps, signal splitters, directional couplers, and line extender amplifiers connects to the bridging ports. A cascade of subscriber tap values in a descending order usually connects to a bridging port. When a signal split is needed, an amplifier may also be required to compensate for the tap through loss and or splitter or directional coupler loss. System design computer programs keep track of the signal level values and indicate when an amplifier is required. Often, the exact placement is made by the program operator. The amplifier location can then be placed on a more convenient pole or pedestal, making maintenance and repair work easier and less hazardous. It is the distribution plant that often provides the greatest number of design choices available. In general, bridging and line extender amplifiers operate at high gain and output levels; consequently, the feeder cascades are limited to two amplifiers.

Figure 2-35
Map symbols for
trunk bridging

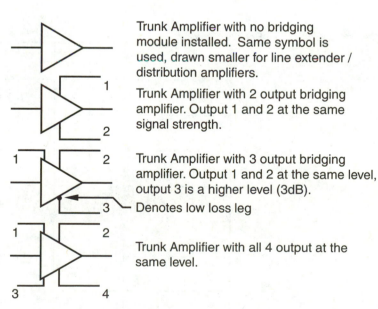

Trunk Amplifier with no bridging module installed. Same symbol is used, drawn smaller for line extender / distribution amplifiers.

Trunk Amplifier with 2 output bridging amplifier. Output 1 and 2 at the same signal strength.

Trunk Amplifier with 3 output bridging amplifier. Output 1 and 2 at the same level, output 3 is a higher level (3dB).
Denotes low loss leg

Trunk Amplifier with all 4 output at the same level.

Reverse Amplifier Considerations—Two-way Services Systems using two-way services have reverse-amplifier and diplex filters installed in the amplifier housings. These amplifiers operate in a reverse direction to the main downstream cable system. A band of frequencies is allocated for the reverse system and diplex filters are used to separate the downstream and upstream frequency bands. The 5- to 40-MHz upstream band is below the 55- to 750-MHz downstream band and is referred to as the subsplit reverse system. When the reverse band is placed in the middle of the down-stream or forward band, it is known as a midsplit system. Placing the reverse band at the upper band edge results in what is termed a high-split system. The most common of the reverse bands is the subsplit system. This narrow band can be used for approximately seven television channels, but the tightness of the system determines how and where the noise build-up occurs in the reverse band and the placement of the reverse carriers within the band.

The noise level in the active reverse band can be measured using a spectrum analyzer. A typical spectrum analyzer display is shown in Figure 2-36. Before using the reverse system, the amplifiers and filters have to be installed. Since the cable loss is much less in the sub-band, it determines whether an amplifier is needed at every station.

Figure 2-36

Noise spectrum of a typical sub-split reverse

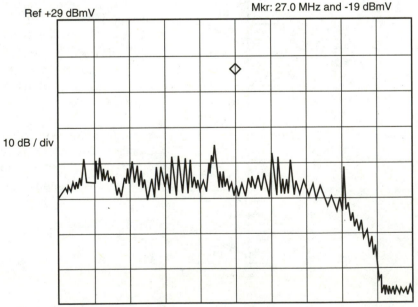

Ref +29 dBmV

Mkr: 27.0 MHz and -19 dBmV

10 dB / div

Start - 2.0 MHz
Resolution 100 KHz

Stop 52 MHz

Essentially, the design and projected performance calculations for each path in the reverse direction should be done. Again, many *computer aided design* (CAD) programs contain sections on reverse system design. The reverse system should be activated on a leg-by-leg basis. When activating a reverse cable section, it will most likely be necessary to tighten or repair any leaky connectors, correct drop problems, and install reverse band stop filters in drops to reduce noise and signal ingress in the reverse band. In short, the system has to be reworked on a section-by-section basis. Now signals can be placed and balanced on the reverse system.

A spectrum analyzer measurement of the return signal can be measured at the head-end. Several manufacturers provide instruments that are especially useful in balancing and monitoring the return signals. It is a constant maintenance problem to keep both the forward system from leaking signals and the return system from signal and noise ingress. Essentially, this means that all connector cables and device housings have to be tight and corrosion-free. Reverse system testing and maintenance topics will be studied in detail in Chapter 6, "Cable Plant Testing and Maintenance."

Reverse or return systems have several areas of application in the cable television industry. One earlier application was to obtain programming from local schools, municipal meetings, and so on, which was done live and sent back to the hub/head-end for downstream broadcasts to the subscribers. After local groups found how much work was involved in producing an hour's programming, interest decreased dramatically. Most subsplit return systems with two to four television channels available are found to be adequate for most community requirements.

Some of the more urban communities required a municipal institutional cable system called an "I" loop. Such systems could be simply a few sections of cable plant operating in both directions connecting several municipal locations together. One usual connecting point was the local cable operator's hub/head-end. An example of such an institutional loop is shown in Figure 2-37. Such a system can be used to carry a full spectrum of signals in each direction carrying video (TV), voice, and/or data on modulated RF carriers. Switching functions at the hub/head-end could broadcast any selected channel(s) over the forward system for subscriber viewing. This type of system was definitely a significant financial burden for the cable operator and, fortunately for most, few such systems were built.

Maintenance of such systems is essentially the same as for the forward system. For wide-band forward and return systems, both a high- and low-pilot carrier are required at the sending ends. These carriers could be modulated with television signals and used as an ordinary channel, but they

Figure 2-37
Institutional cable
system overlayed on
subscriber cable
system

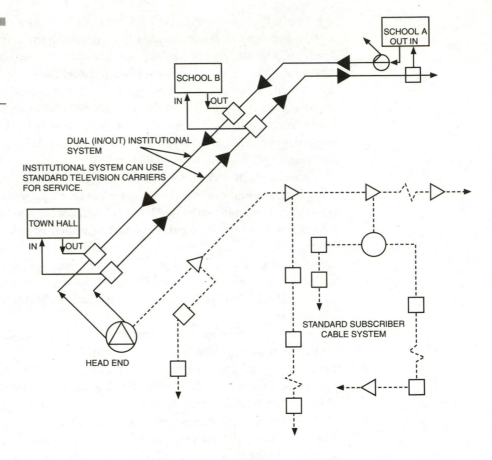

have to remain on the system used or unused as the pilot carrier reference needed to maintain proper system gain and slope.

Construction Practices

Construction practices and construction equipment have definitely improved since the early days of cable television systems. The method for servicing an aerial plant used to be climbing the poles with a set of hooks and a belt. Most people learned to climb poles from earlier experiences working for either the telephone or electric company. Some learned to climb from a tour in the service, usually the U.S. Army Signal Corps. Essentially, the Signal Corps adopted telephone company methods and specifications. Therefore, former Signal Corps veterans often became employed by the cable television industry companies and learned the technical side during the early start-up years. Those that did not climb often used extension ladders against the poles.

Some innovative cable system technicians experimented with vehicle-mounted ladders. In a short time, the hydraulic lift truck became available and soon became affordable and practical for cable television system use. As suburban areas grew and became more populated, underground utilities including cable television systems became more attractive. Many system operators felt their plant was safely underground and at ground level for easy maintenance. Smaller and more affordable trenching machines made such installation financially viable.

Aerial Construction Methods   Aerial construction methods have definitely undergone evolutionary changes. From a labor-intensive manual operation to modern hydraulically operated lift trucks with hydraulically operated drills, saws, and hoists, present-day construction practices have drastically improved. Some old-timers remember climbing the poles and boring through a pole with a bit brace and ship's auger. Nowadays a person is lifted up the pole in an aerial basket where the pole is drilled in about a minute and the through-bolt washers and nut are installed in another minute, enabling many more poles to be framed in a day. Such efficiency did not result in many cost reductions since the equipment costs had to be repaid and the operators paid more for being more experienced, but a lot more plants could be built in a shorter time frame.

Pole line hardware also improved over the years and more attention was paid to the effects of dissimilar metals on corrosion. Therefore, better materials were introduced. Usage of corrosion-resistant materials such as a Monel metal, stainless steel, and a variety of plastics produced a more reliable cable plant.

Underground Construction Methods   Underground cable plant construction in both methods and materials has changed more than aerial plant. First, the method of placing the plant underground has certainly progressed from "pick and shovel" methodology. Opening a trench has changed from a back-hoe machine to a chain-type trencher that opens a narrow trench, spilling the removed earth into a neat row along one side of the trench. Usually stones are manually removed and often a layer of gravel is placed in the bottom of the trench. A trench of this variety can be used with direct burial cable. Often PVC-rigid conduit is placed in the trench with an installed pull-rope to facilitate pulling the communication cable through at some later time. Semi-flexible conduit is placed in the trench and some manufacturers provide the conduit with the cable already installed. Several conduits can often be placed in the same trench. Many trenchers also have an optional back-fill blade that can plow the spill back into the trench. Trenching machines can provide trenches from very shallow to several feet deep. For most communication purposes, a depth of 24 to 30 inches is appropriate. A favorite method of placing cable underground is using a cable plow. As mentioned in Chapter 1, "Introduction," the two types of cable plows are the vibratory plow and the stationary plow.

Since an underground plant requires access to the active and passive devices, a variety of equipment enclosures are designed for underground plant servicing. These enclosures fall into two categories: aboveground and underground. The aboveground enclosures are either metal or plastic pedestals, and the underground enclosures are either plastic or concrete vaults. The vaults are usually fitted with an appropriate cover that either has to be pried open with a special tool or unbolted with a special wrench. The aboveground pedestals have either a locking device or special tool. These measures prevent unauthorized tampering with the equipment. More progressive cable operators that have electronic monitoring capabilities, called status monitoring, can integrate an alarm system for vault or pedestal protection. Some of these pedestals and vaults are shown in Figure 2-38.

Figure 2-38
Types of pedestals and vaults

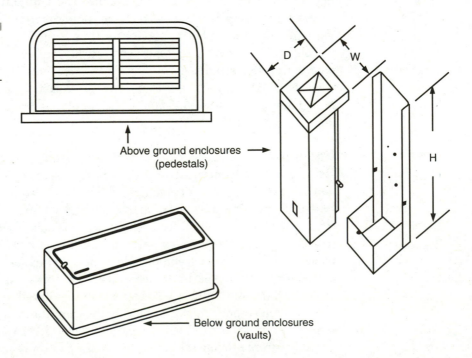

Above ground enclosures
(pedestals)

Below ground enclosures
(vaults)

Underground cable plants have to be mapped and properly marked so that service and maintenance can be done at a later time. Unlike an aerial plant, an underground or buried plant, as it is often called, cannot be seen. Therefore, the exact location of cables has to be known and appropriately mapped. Most of the problems with a buried plant result from the careless cutting and digging by other underground utilities. The installation of traffic signs often causes service interruptions for the cable, power, and telephone operators. Thus, various measures are employed to prevent damage to an underground cable plant. The installation of a yellow or orange plastic warning tape placed over the cable a few inches below the ground's surface offers some protection against digging up the cable. Warning decals placed on the pedestals can also alert others that facilities are buried nearby.

Cable operators with an underground cable plant should have some type of cable *Time Domain Reflectometer* (TDR) available to find the distance from one point to the fault. Making a measurement from each end of a cable with a fault (a cut or a crush) will give a close location of the fault. Also, cable operators need some type of cable location instrument, of which several types are commercially available. Essentially, they place an electrical signal on the cable that is detected aboveground by a receiver. In most cases, the transmitted signal is placed on the cable's metal outer sheath at one end of the cable run and a person with the receiver walks along the cable run with the receiver, detecting the path of the cable by marking it with either spray paint or garden lime.

Once the distance to the fault has been determined and the path of the cable is found, the close location of the fault is identified. Following this procedure minimizes excessive digging and makes repair of the problem much faster. This method is illustrated in Figure 2-39. Underground plants are becoming more and more attractive to cable operators simply for the reasons of maintenance and safety from vehicle accidents.

Figure 2-39
Location of buried cable path

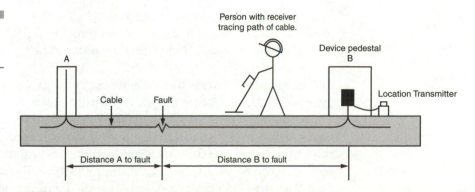

Splicing and Connectorizing Cable connectors have also certainly improved over the years. One problem with connectors is what is known as a pull-out, in which the cable is simply pulled out of the connector, causing a loss of signal and cable power. This is the most catastrophic problem with a connector.

Another problem is corrosion and connector deterioration over time. Systems with such problems will be faced with accompanying outages and subscriber complaints. Since we live in the time of corporate mergers, several connector companies have been bought out and or consolidated. More attention is consequently being paid to the metallurgy involved with cable plants and the problems of mixing various metals. Certain metals when in contact with one another in a moist and corrosive climate (such as air pollution or acid rain) can cause rapid deterioration of the plant. Also certain metals used together in housing and fastenings can cause threads to seize. Different metals have different thermal characteristics that can cause expansion and contraction problems over seasonal temperature variations. Fortunately, several excellent companies still provide high-quality connectors to the cable television industry.

Another consideration cable operators have to make is the ease of connectorizing the system, which means the ease of preparing the cable and installing the connectors. Proper tooling is a necessary requirement. Present-day connectors require that some of the dielectric has to be removed so the internal R.F.I. sleeve can fit under the aluminum sheath. This operation is known as coring and can be done with a manual corer or one fitted on an electric drill. The drills used by cable splicers when connectorizing a plant can be powered by electricity (portable), compressed air, or by hydraulic means. Use of powered coring methods speeds up the connectorizing process. Once a connector manufacturer is chosen, the application's engineering department and/or customer service department will usually be most helpful to the cable operator, providing installation and tooling information. Most splicing contractors are familiar with all the available connectors and the installation methods required. Proper installation is most important in keeping the power flowing and the signal inside the cable.

Coaxial cable systems have to conform to the FCC's required signal leakage specifications, testing the plant several times a year for signal leakage. Connectors that prevent signal leakage also shield the plant against extraneous signal ingress and noise ingress. Improper connectors do not give a good impedance match, which can result in standing waves and signal echoes, which result in ghosting of the TV picture. By now, it should be evident that good connectors properly installed will provide good cable plant operation over the years. Some of the cable preparation tools appeared in Figure 1-6 in Chapter 1.

Grounding and Bonding Requirements   Proper system grounding and bonding is extremely important to protect cable systems from the effects of lightning and power system surges. For aerial plants, the practice is to bond at the first and every 10th pole, as well as the last pole in a cable run. Some pole attachment agreements call for different grounding and bonding practices that must be followed. In most cases, all power supply locations for either aerial or underground plants must have a separate ground rod. Ground wires on utility poles often require the installation of protective pole molding to protect the ground connections. Special bonding clamps are used to bond copper wire, usually number 6 (AWG) soft-drawn copper wire, to the galvanized steel messenger strand supporting the aerial cable system. These clamps are usually plated brass. Some aerial bonding methods are shown in Figure 2-40.

Grounding and bonding at the head-end and/or hub distribution points is crucial for protecting sensitive electronic equipment from lightning and power surges. Presently available are surge protectors that can be wired into the power entrance panels of hub/head-end facilities. Such devices can protect all the equipment inside the building that is connected to the inside power distribution system. Emergency stand-by power systems of the external engine-powered variety also have to be properly grounded and bonded to the central grounded electrode. Essentially, the main ground electrode is connected to a system of driven ground rods and the power entrance ground. A hub/head-end ground and bonding plan is shown in Figure 2-41.

Subscriber Equipment and Signal Security

Subscriber- or customer-used equipment provided by the cable operator is essential to the delivery of a high-quality product. This equipment should be user-friendly and attractive in the television viewing system since it appears near or on top of the subscriber's TV set. Unfortunately, the use of a set-top cable television converter makes the subscriber's TV set remote control quite useless since channel selection is done by the set-top cable converter, as discussed earlier. This also makes VCR recording for the subscriber quite difficult and compounds the problem of the VCR remote control. Signal security in most cable systems is performed within the cable converter where the premium channels are decoded or descrambled.

Figure 2-40
Example of an aerial
plant grounding and
bonding

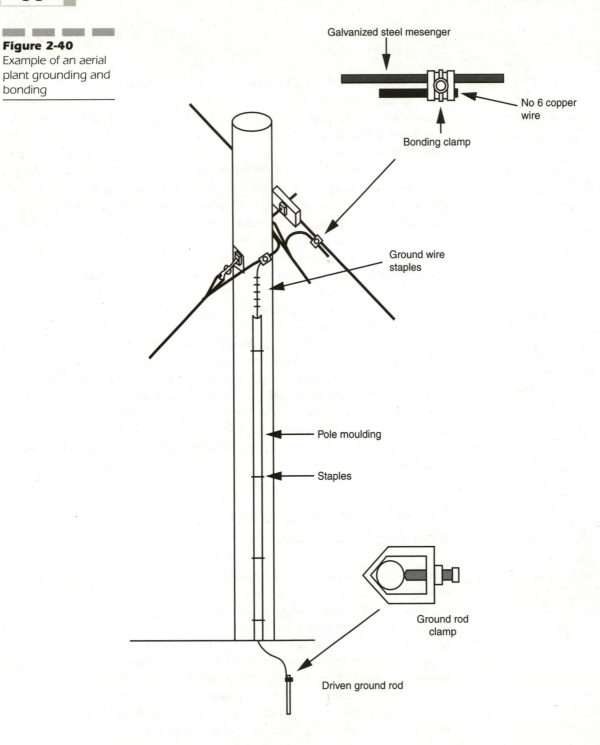

Galvanized steel mesenger

No 6 copper wire

Bonding clamp

Ground wire staples

Pole moulding

Staples

Ground rod clamp

Driven ground rod

Figure 2-41

Example of a head
end ground/bond
system

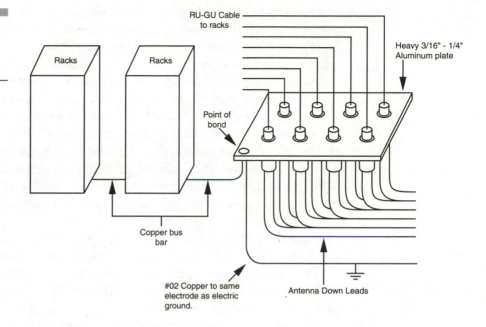

The Set-top Converter   The set-top converter has evolved over the years from the simple mid-band to the super-band channel selector. Today most cable converters still perform channel selection and operate with their own remote control. Decoding and descrambling takes place within the converter for the premium pay channels.

Ordering premium channel programming is still mostly done by a phone call to the cable operator's subscriber office. In many instances, cable operators use what is known as addressable converters, and when a subscriber calls the cable office and orders a premium channel or a pay-per-view program, the converter is authorized by a downstream data channel. This authorization activates the converter to decode the premium programming ordered. Cable systems with active reverse systems can place the ordering requests through the bidirectional set-top converter to the hub point. The subscriber can now order a particular pay-per-view program by using the set-top remote control to place the order. This type of television service is enough to make us all couch potatoes.

Positive/Negative Signal Trapping   Prior to such sophisticated converters, signal security was provided by what is known as signal trapping. The unsubscribed premium program channels were trapped out (prevented) of the subscriber's drop cable. This was done at the tap port feeding the subscriber's drop cable.

Trapping devices appear as one of two types. A device that prevents the premium signal from entering the subscriber's drop is essentially a narrow-band stop filter, known as a negative trap. The other trapping device is what is known as a positive trap. At the hub/head-end, an interfering signal is placed on the premium channel that renders the channel unwatchable. The positive trap removes the interfering signal from the premium channel, now rendering it watchable. All subscribers not wishing to have the premium channel will need a negative trap, while a subscriber wanting the premium channel will need a positive trap. Systems that have a negative trap scheme with few subscribers taking the premium service will need a lot of negative traps. For such systems, a positive trap scheme will be more cost-effective since the few premium program takers will need the interfering signal removal traps.

Since these traps appear at the tap port, a system accounting (audit) of subscribers taking or not taking the premium service can be made by a simple drive-by for observing the aerial traps. For underground systems where the taps are in pedestals, the pedestals should be locked to prevent tampering with the traps. Removing a negative trap allows the premium program to be seen. A positive trap of the proper type is needed to remove the interfering signal for the premium channel to be seen. Therefore, signal security is better using the positive trap method since an appropriate device in the drop has to be in place in order to receive the premium signal. Unfortunately, unscrupulous people sell such devices to those people who wish to steal the service that are usually inserted within the TV set where it is unseen. Some people use dummy negative trapping filters at the tap port that look like a trap but just pass the signal through.

Scrambling/Descrambling Scrambling or encrypting a television signal, rendering it unwatchable, is indeed a better and more secure method to prevent stealing or unauthorized signal use. This is a more elaborate and expensive method, but many cable operators feel it is worth the expense.

Several methods are used to scramble or encode the premium channels. One method is to remove the horizontal synchronizing signal, commonly known as sync suppression. This type of scrambling is, of course, analog and works well with the standard NTSC video format. Inverting the horizontal synchronizing signal with some form of alternating pattern offers a higher level of scrambling. Such methods working in conjunction with subscriber terminal addressability are still being used by many cable operators. With the arrival of digital television, digital encryption techniques will certainly improve signal security to a very high level. This apparently is what the cable television industry has been waiting for.

Addressable Converters   Addressable converters, or as they are known, addressable set tops, do two things for the cable operator. The most important consideration is that subscribers can order program services over the telephone and the cable operator can then download the authorization codes to the set-top converter. Thus, the converter is activated to automatically make the changes and descramble any of the selected premium services. Addressability enhances signal security since the set-top has its own address code where its location address is known. Now if this box is activated and stolen, and the service is not paid for, the box can be rendered inactive and hence useless.

Interactive Converters   Interactive converters are those set-tops that can communicate directly with the hub/head-end. Such converters make pay-per-view channels quite easy to be seen by any of the subscribers and services can be ordered through the remote control directly. These converters can work the *video on demand* (VOD) systems some cable operators presently have in place. Converters operating in this manner must have a reverse path of communications back to the main signal distribution point. This reverse or return path usually operates in the sub-band (sub-split reverse) portion of the cable system. A relatively small number of cable systems have the return path in the mid-band or high-band that actually contains a wider return band, offering more equivalent upstream television channels. Interactive converters can be traced if stolen and detected in the wrong part of a cable system.

Connecting VCRs, TV Sets and Converters   Connecting subscriber equipment to the cable system probably causes the most trouble for the subscriber. It is basically the set-top converter that causes the television set and VCR remote control to be ineffective in selecting television channels. This problem causes more complaints from subscribers with cable service. The use of switches to change or reconfigure the subscriber's TV-VCR hook-ups may help the situation. Also smart remotes (programmable remote controls) can be programmed to operate all of the subscriber devices.

This problem really resulted from the television manufacturer's inability to address the cable television industry's specifications. A possible solution would be for the TV set to operate at the intermediate frequency provided by the set-top box and the VCR. This subject has been under much discussion between cable operators and the consumer electronic manufacturers with the Cable Labs and the FCC advising at times. Several subscriber equipment connections are shown in Figure 2-42.

Figure 2-42
Adding another
converter, T.V. set and
VCR can make for a
complicated home
distribution system

ConfigurationA: TV Set 1 and VCR views the unscrambled channel tunned by the converter. TV Set 2 simply views any clear channel, if TV Set 2 is cable ready.

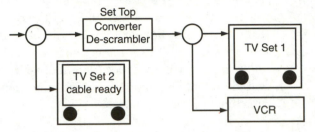

ConfigurationB: TV Set 1 can view any clear cable channel when A/B switch is in position A. VCR can be recording a pay TV channel selected by the converter. For the TV set to view the pay channel, switch to postion B and tune the set to the converter output channel.

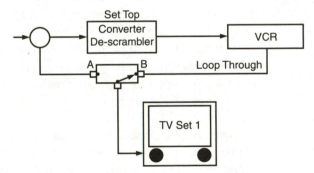

The *National Cable Television Association* (NCTA) released a publication addressing the interconnections between the cable converter, the TV, and the VCR. This publication covered a large variety of interconnections using A-B switches, splitters and couplers, multiple converters, VCRs, and TV sets. Many homes subscribing to a cable service that are supplemented by a *Direct Satellite Broadcast System* (DBS) as well as a VCR can have an elaborate switching distribution system.

A word of caution: when a subscriber has an off-air roof-top antenna connected to the switching coupling network, care should be taken to prevent the wide-band cable system from being connected to the antenna. This could cause some serious interference problems with many homes in the neighborhood as well as the cable operator. Remote as it may seem, it can happen and it has happened before. Such signals can disrupt air traffic communications as well as some of the law enforcement and safety mobile radio services. Another word of caution is to make sure switching and coupling devices are the appropriate ones to use for cable television use. Some available devices are made for base-band video and are not appropriate for higher RF frequencies.

Probably one of the most important considerations with subscriber-connected equipment is the connectors themselves. Some of these connectors are used by the subscriber and are obtained at local electronic outlets, hardware stores, and homeowners' outlet stores. Unapproved tools and techniques are often used and hence the connectors are improperly installed. This is beyond the control of the cable operator who only learns about it through resulting service calls reporting poor signal problems by the subscriber.

The connector used to connect subscriber equipment is the "F"-type connector, which had its infancy in the MATV industry. This connector evolved from a rudimentary connector with a separate one-eighth-inch crimp ring to a single-piece weather-tight connector that requires proper stripping, tooling, and a more precise crimp tool. Several manufacturers supply versions of these superior type "F" connectors to the industry. Most cable operators use this type of connector and have instructed their installer personnel in the proper installation procedures. However, a lot of the old connectors are still out there in many cable systems.

Changing these connectors is an expensive and labor-intensive procedure and is usually done during a system upgrade. These poor, early "F" connectors cause signal leakage, ingress, and noise problems that interfere with the use of an active return signal plant. When a cable system is planning a system upgrade requiring an activated return path, changing all the subscriber "F" connectors on an area-by-area basis is necessary in order to control the noise and signal ingress in the return path. Often subscribers not using the return path will have a filter trap at the return band frequencies, which will prevent signals or noise entering the return path from this location.

In many instances, cable systems discover signal leakage from a subscriber address during the usual periodic leakage testing required under the FCC rules. The usual procedure is to disconnect the subscriber drop and check the signal leakage level. If it goes away, then the subscriber will have to be contacted so an appointment can be made to make the necessary corrections. Since all of these costs have to be born by the cable operator, many cable systems avoid return path activation. An example of a current "F"-type connector and the accompanying cable preparation is shown in Figure 2-43. It should be obvious that this is indeed an improved connector that can be used with single and multishielded drop cables.

Figure 2-43
Example of a
currently used "F"
connector stripping
and crimping

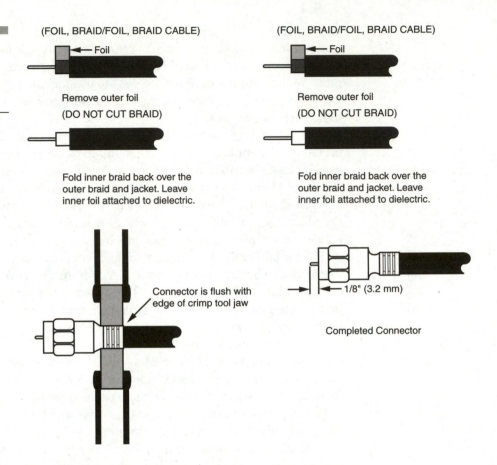

(FOIL, BRAID/FOIL, BRAID CABLE)

Foil

Remove outer foil
(DO NOT CUT BRAID)

Fold inner braid back over the
outer braid and jacket. Leave
inner foil attached to dielectric.

(FOIL, BRAID/FOIL, BRAID CABLE)

Foil

Remove outer foil
(DO NOT CUT BRAID)

Fold inner braid back over the
outer braid and jacket. Leave
inner foil attached to dielectric.

Connector is flush with
edge of crimp tool jaw

1/8" (3.2 mm)

Completed Connector

System Testing/Proof of Performance

Cable television systems have to be completely tested on an annual basis for proper system performance parameters as required by the FCC under Part 76 Title 47 of the Federal Commission Rules of 1990. The first test period should be when the system is first turned on after construction is completed. A system operator should insist that the new system pass all FCC tests satisfactorily before final acceptance and the contractors are paid. In short, it should be a part of the construction contract in the final check-out and acceptance.

For the first three years, annual proof of performance tests are conducted on an RF basis, after which tests are performed on base-band signals on a three-year cycle thereafter. RF tests take place twice yearly and base-band tests are every three years. Systems with fewer than 1,000 subscribers are still exempt from proof of performance testing. Such systems consist of neighborhood systems similar to MATV systems and evidently seem trivial to the commission. Systems with 1,000 to 12,500 subscribers have to be tested at six widely separated test points where at least one-third must be set at the most distant point from the central signal source. For systems greater than 12,500 subs, another test point needs to be added for every additional 12,500 subscribers.

Each system has to have in its company technical records a copy of the FCC rules and Regulations Part 76 Title 47. Test records have to be kept on file for a five-year period and are subject to inspection at any time by a representative of the commission. The test results also must include the test procedures, calibration data and dates for the test instruments used, and the qualifications of the persons conducting the tests. The test points are specified at the subscriber terminal, which in most systems is the converter output terminal. Since most converters are pre-tested, tests can be made at the end of a 100-foot drop from the system tap and can be corrected by the converter performance specifications.

Initial Performance and Testing and Balancing   Initial system test results are important in setting the system performance standards. In most instances, a newly tested system never performs better than at turn-on time since system component and cable aging causes a gradual system deterioration. A properly maintained and operated system periodically tests and services a plant to maintain proper signal quality. This, of course, is just good business, giving quality service to customers.

The turn-on sequence of events is important when performing initial tests. The procedure is to turn-on plant distribution power, correcting any open or short circuit problems. Next, the pilot signals are injected into the system, and amplifier modules and power modules are installed in the trunk housings. At this time, each amplifier is roughly balanced on both the high and low pilots. Following this operation, a leakage test of the trunk runs is made so any further balancing will not compensate for a poorly impedance-matched plant. At the initial leakage test phase, the whole system can be swept using a sweep generator receiver instrument. This can be done on a system segment or a main trunk branch at a time. Lastly, the whole signal spectrum can be turned on and the system can undergo the final balance and leakage test. The results should be kept in the system file for future reference if any troubles should occur.

Often the costs of final testing are the responsibility of the prime contractor when the results prove to the system operators that the system is complete and operational according to the construction contract. Lending institutions holding the loan on the system may require either a consulting engineer as an observer of the final testing program or an independent testing company to perform the test to the satisfaction of all concerned parties.

Signal End Level Testing   The next most important test program is the periodic signal-level measurements of end points in the system. Unfortunately, many systems rely on the sporadic signal-level measurements made by drop installers or maintenance technicians making trouble calls. The measurement of system end levels on all cable channels can give a lot of information about how the system is operating. Seasonal variations of a signal level can indicate how well the thermal compensation and AGC/ASC systems are working. An analysis of this data either by a manual method or a computer program can indicate areas of the plant that need some maintenance long before the phones start ringing with trouble calls. Inexpensive graphics software can be used to present spectrum plots at the end points of the system.

For an aerial plant, a drop can be made at a tap fed from the last line extender and stapled down the pole. On the end of the cable, a connector installed with an F-81 barrel can provide a convenient point for a technician to connect a signal-level meter lead. The whole drop test point can be hidden behind pole molding. At the end of the test, the technician can remove the instrument lead and install a 75-ohm terminator to the test drop. Many instruments can provide the whole spectrum on an LCD display and many can store data in a solid-state memory. End-level tests can be made either on a monthly or bimonthly basis, or at least on a quarterly plan, so data can be accumulated over a significant temperature range.

Leakage & Ingress Testing CLI   Another routine test that cable systems must peform is the signal leakage test required by the FCC. This test has to be done on a quarterly basis and the results must be kept in the company technical files. The *Cumulative Leakage Index* (CLI) parameter will be computed for a system at the completion of the testing period. Any problems with signal leakage should be expeditiously repaired with documentation on the cause, the fix, the date, and the personnel. The methods and technical specifications will be discussed later in Chapter 6.

Two basic test procedures can be used to obtain the data for the CLI. One is the fly-over method, requiring the use of an airplane flying over the system at a fixed altitude. Monitoring equipment onboard the aircraft will record the signals leaking from the cable plant. Once the signal levels and the location are known, the problem areas can be easily identified. The use of the satellite-based *Global Positioning System* (GPS) can also be a big help in plotting the location of leaking plant. The second method employs vehicle-mounted antennas used to locate problem areas. Once these areas are found, technicians using probe antennas can pinpoint the faulty connectors and loose amplifier housing. Many cable systems use this method. Technicians doing installation and maintenance tasks can monitor the plant at the same time. Keeping a plant free of leakage will also keep it free of signal noise and ingress.

Summary of Testing   This chapter essentially addressed coaxial-based cable television systems. Information in this chapter supported the more elementary nature of the first book, *Cable Television Technology and Operations*. Coaxial cable plants developed from the early rudimentary systems to very wide bandwidth coaxial cable distribution systems, carrying a multitude of programming covering entertainment, news, weather, sports, education, and community services. Systems with long trunk runs operating at 400 to 450 MHz upper limit had to use power-doubling amplifiers and/or feed-forward types to reach the ends of the system. The placement of nodes or distribution points fed from super trunks was one method used to limit cascaded amplifier problems. This type of cable plant employed a large number of amplifiers (actives) along with signal couplers and dividers as well as the required number of connectors.

Problems with system leakage, noise, signal distortions, and high power consumption made for very expensive plant requiring a lot of maintenance. Microwave radio links were also employed to share head-end and satellite receiving systems with a number of signal distribution points. Technical personnel had to have all the necessary technical skills needed to service and maintain a variety of equipment. And along came fiber optics, which for many systems was just what was needed. For most systems, replacing the trunk system with optical fiber shortened amplifier cascades and provided a superior signal to the signal distribution nodes. System leakage was reduced and power consumption was reduced as well as amplifier cascade noise and distortion components. Technicians now found that more training was required to learn the installation, maintenance, and testing procedures for the optical fiber section of the cable television system. The next chapter is a study in fiber-optic communications.

Fiber-Optic Technology in Cable Television Systems

Introduction

Fiber-optic technology as applied to cable television systems was just what was needed and at the right time. Cable systems were expanding their bandwidth with the upper limit approaching one GHz. The number of amplifiers per cable-mile of a plant was increasing at an alarming rate. High power consumption required more power supplies, which in turn caused system costs to increase. More cable plant devices and connectors caused increased signal leakage problems, all contributing to increased plant maintenance and escalating costs. It was indeed a difficult and expensive project to expand plant bandwidth and channel capacity. Fiber-optic technology started out as simply plastic light pipes acting as monitor indicators for a variety of applications including automobiles. The continued research into the field of optics and optical fibers finally produced fiber-optic communications.

Fiber-Optic Development

The development of optical fiber technology began in the 1950s and 1960s with the invention of the Light Amplification by Stimulated Emission Radiography device, the laser. Early lasers were essentially research tools used to study the transmission of light through various optical devices, glass fiber included. The first lasers were gas-type lasers with large amounts of light energy transmitted. The manufacturing methods used to form glass (silica) fibers were developed by companies here and abroad. As light sources were developed, glass fibers contributed to the development of fiber-optic communications.

History of Fiber Optics The significance of the laser was due to its capability to supply a high-intensity light output at essentially one wavelength and be coherent (in phase) as well. Thus, it was described by its action as a monochromatic coherent light source. The laser of choice for communication application became the solid-state laser diode. This device could be made small enough to be mated with the end of an optical fiber. The development of the photodiode was also important to the communication application because this was the photo detector or receiver. This photodiode was the transducer that converted the received light energy to electrical energy. Now the ingredients were available for a fiber-optic communication system. The transmitter, the fiber-optic cable, and the receiver were now available.

Optical Physics Review   A review of applicable optical physics will help give more of an insight into how an optical fiber communications system operates. For a fiber-optic communications link, the carrier is optical energy that propagates through the fiber to the receive point. The optical transmitter at the sending end has to be modulated with the electrical signal. At the receiving end, the receiver converts the modulated optic wave to electrical energy that is nearly an exact replica of the sending end electrical signal. The transmission medium from the sending end to the receiving end is a glass fiber.

Two types of silica glass fibers exist: one is called multimode and the other single mode. The modes are essentially the same as waveguide modes in the RF domain.

We should recall that radio frequency (electromagnetic) radiation travels at the speed of light (in a vacuum and nearly so for air). This is because light energy is electromagnetic energy and propagates through space as well as glass fibers. It should come as no shock that glass fibers operate as waveguides for light energy. Often when light energy is mentioned, we think of visual light. Also, the word optical often refers to vision. If a person looks through a thick sheet of window glass edgewise, the light generally looks greenish and is reduced in brightness. Therefore, this type of glass would not work very well if used to form a fiber. In brief, the glass has to be extremely clear and transparent to light energy. Glass manufacturers discovered through research that transparency is a function of color or wavelength. Since optical energy is electromagnetic, a spectrum of electromagnetic energy is shown in Figure 3-1.

It should be noted that the visual light spectrum with its associated colors occupies an extremely small portion of the frequency spectrum. The plot of an optical silica glass fiber is illustrated in Figure 3-2. Notice that the visible spectrum is to the left of the graph (less than 1,000 nm). Also notice for a wavelength of 1,550 nm and 1,310 nm the loss is 0.2 dB/km and 0.3 dB/km. This simply means that energy at 1,550 nm of a wavelength will be decreased only 0.2 dB for the length of one kilometer. A coaxial cable with one of the lowest losses of 1,000 MHz is about 1.3 dB/100 feet at 68° F. This relates to approximately 43 dB/km. It should be clear that it is this an extremely low loss of optical fiber, making its use attractive in the communications industry.

Types of Optical Fibers   The first generation of optical fibers was little more than light pipes used as indicating devices such as automotive lamp monitors and inspection monitoring equipment in medicine and industry. As research and development continued, it was found that if the glass fiber was coated with glass of a slightly lower refractive index, total internal reflection would take place. Thus, light energy entering the fiber would stay in the core and travel to the end of the fiber. This outer covering of glass over the core is known as the cladding. Light rays essentially propagated through the fiber core in waveguide fashion, following a multitude of mode paths, and multimode optical fiber resulted.

Figure 3-1
Electromagnetic spectrum

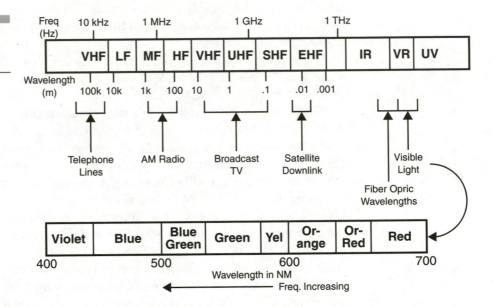

Figure 3-2
Attenuation vs. Wavelength for a glass single mode fiber

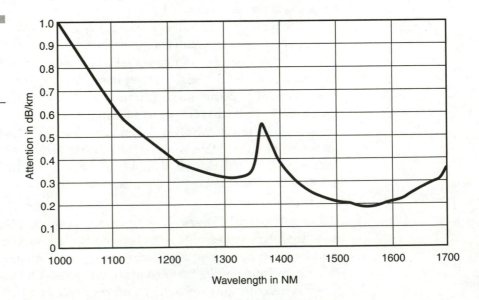

Variations in the geometry of the core/cladding cross-section at the end of a fiber results in several sizes. When the refractive index of the cladding is slightly less than that of the core and the change is abrupt, the resulting optical fiber is known as the step index. This means that there is a step change in the refractive index at the core cladding boundary and such optical fibers are known as step index multimode fibers. This type of optical fiber is shown in Figure 3-3. Rays of light energy propagate through the fiber following different paths due to the different modes of operation, as shown in Figure 3-4.

Due to the different path lengths, light rays at the end of a fiber arrive at different times. This causes a sharp input light pulse to become stretched in time at the output. This stretching in time is known as modal dispersion. Modal dispersion can become a limiting factor for a high pulse rate found in digital signal transmission.

Early fiber-optic systems were used in *local area networks* (LAN) with data (digital) communications connecting computer workstations to printers and file servers. Since many of these systems did not consist of long fiber runs or excessively high pulse (data) rates, system performance was adequate. As system requirements called for greater distances and higher data transfer rates, modal dispersion became a limiting factor. The industry

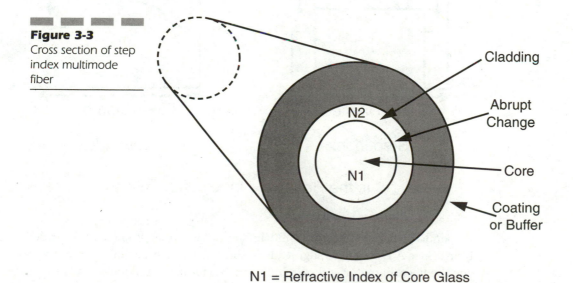

Figure 3-3
Cross section of step index multimode fiber

Cladding

Abrupt
Change

Core

Coating
or Buffer

N2

N1

N1 = Refractive Index of Core Glass
N2 = Refractive Index of Cladding Glass
N1 > N2 by Approx 1%

Figure 3-4
Multimode light wave
propogation through
fiber

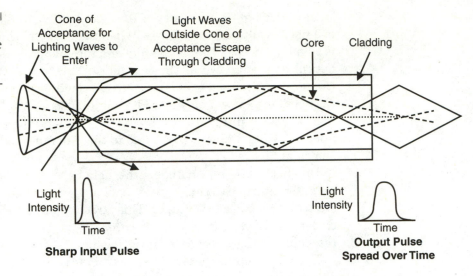

Cone of
Acceptance for
Lighting Waves to
Enter

Light Waves
Outside Cone of
Acceptance Escape
Through Cladding

Core   Cladding

Light
Intensity

Time

Sharp Input Pulse

Light
Intensity

Time

**Output Pulse
Spread Over Time**

Note: Multimode Progation Causes Output Pulse Dispersion

Figure 3-5
Refractive index of
types of optical fibers

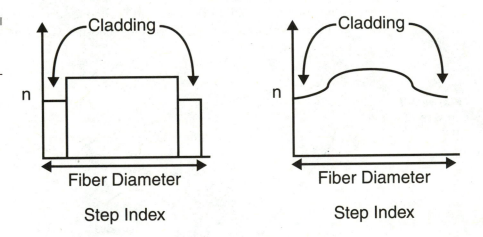

Cladding

n

Fiber Diameter

Step Index

Cladding

n

Fiber Diameter

Step Index

"n" Is the Refractive Index of the Glass

responded with an improved multimode optical fiber known as a graded
index fiber. A plot of the magnitude of the refractive index of the glass versus the fiber diameter is shown in Figure 3-5 for both step index and graded
index multimode fibers.

Notice in the figure that the refractive index of the glass for the graded
index optical fiber decreases gradually (graded) from the core center to the

outside cladding. Fiber manufactured in this manner allows the rays of light entering the fiber to propagate through the fiber, as shown in Figure 3-6. Light rays traveling in the cladding area where the refractive index is less than the core center travel faster. Therefore, they may travel a longer path yet travel faster to arrive at the end along with rays traveling different paths. The effects on pulsed signals are not as severe as with the step index multimode fiber.

In general, multimode fibers have a relationship between the refractive index of the core and cladding glass and the numerical aperture, as well as the acceptance angle. This relationship is illustrated in Figure 3-7 for a step index multimode fiber.

For cable television purposes, long distances of fiber operating at a wide bandwidth is required and neither of the multimode fibers are applicable. Most multimode systems can operate quite well at shorter distances using LED technology as the transmitter and a photodiode as the receiver. The larger core diameter with the larger numerical apertures allows the use of these diodes. The fiber of choice for high-speed wide-band fiber systems requires the development of a single-mode optical fiber. This fiber has a very small core diameter and a thick cladding surrounding it. The small core diameter makes it more difficult to get much optical energy into the fiber.

When the solid-state laser chip was developed, it made it possible to use single-mode fiber systems. The laser chips were made small and the light-emitting area was matched closely with the end of the fiber. The high-intensity narrow-band coherent laser light source was just the component needed for single-mode fiber operation. The most common single-mode fiber

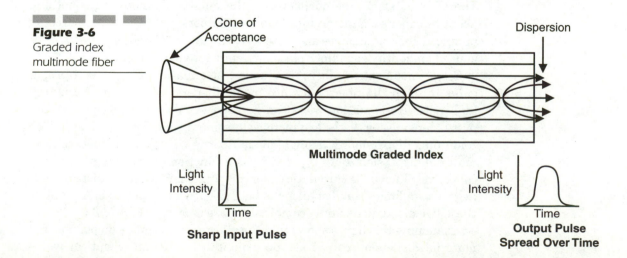

Figure 3-6
Graded index
multimode fiber

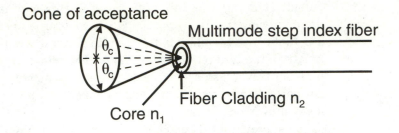

Figure 3-7
Fiber-optic geometry

$$n = \frac{\text{Speed of light in a vacumn}}{\text{Speed of light in the glass}}$$

c = speed of light in a vacumn = 3×10^8 m/s

n_1 - Refractive index of the core

n_2 - Refractive index of the cladding

θ_c - Acceptance Angle

$n_1 > n_2$

Numeric Aperture = $\text{Sin } \theta_c = \sqrt{n_1^2 - n_2^2}$

used today is step index and has a very small core cross section. Figure 3-8 compares the cross-sectional areas for multimode and single-mode fibers.

For the single mode fiber shown in the figure, it should be noted that the outside diameter is the same as for the graded index multimode fiber. Therefore, it should be obvious that the outside diameter says nothing about the fiber's characteristics. The actual fibers making up commercially produced fiber optic cables are coated with a plastic colored covering. A color is used to identify each fiber placed in a protective buffer tube.

Single-mode optical fibers provide low attenuation of light as it propagates through the fiber, and since modal dispersion is not a factor, very high pulse transmission rates can be used. As the upper frequency and pulse transmission rates are increased, some dispersion does result. The remaining dispersion in a single-mode fiber is waveguide dispersion and material dispersion. Both types vary as a function of wavelength and are collectively known as chromatic (color) dispersion. Waveguide dispersion is negative and material dispersion is both negative and positive. A graph of these two dispersions and a combination are shown in Figure 3-9.

Examining this figure shows that the effects of waveguide dispersion and material dispersion cancel at approximately 1,310 nm of optical wave-

Figure 3-8
Fiber crossectionals

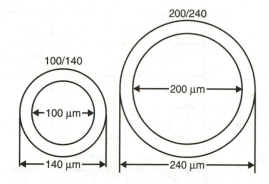

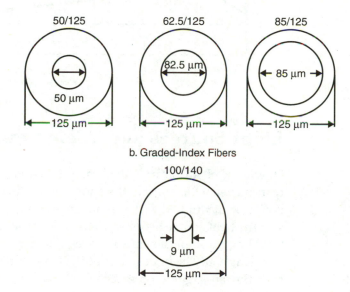

a. Step-Index Multimode Fibers

b. Graded-Index Fibers

c. Step-Index Single-Mode Fiber

length. This results in zero dispersion, so little or no pulse stretching (distortion) results and high optical pulse rates can be used. From examining Figure 3-2, it should be noted that the fiber loss at 1,310 nm is not at the lowest point. The loss is at its lowest at a 1,550-nm area. From Figure 3-9, we can see the dispersion is not zero at 1,550, which means high data rate pulsed signals will be affected. The development of doped fiber results in shifting the zero dispersion point into the 1,550-nm region. Now high data rates at the lowest loss area for this doped fiber become possible. This is the fiber of choice for long-haul optical cable systems such as submarine cables.

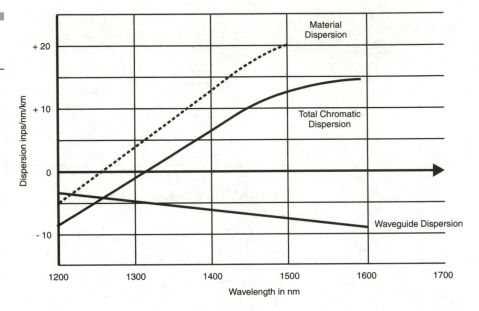

Figure 3-9
Chromatic dispersion
vs. wavelength

Light Sources and Their Development

Operation of an optical fiber communication system starts at the light transmitter, which injects the optical signal into the fiber. This light source has to be modulated with the message signal. The stronger or more intensive the light source becomes, the stronger the received signal. Light energy traveling through the optical waveguide carries the message traffic to the receiving end. The light source at the transmitting end has to be modulated with the message-carrying signal.

The light sources used in current systems are either the solid-state *light-emitting diodes* (LED) or a solid-state laser diode. LEDs are often used for multimode data-carrying LANs and provide satisfactory operations for a variety of applications. The solid-state laser diode is used mainly in single-mode applications, requiring communication distances of 10 to 50 kilometers. This is the type used in cable television systems. There are two basic types of laser diodes: the Fabry-Perot and the *Distributed Feed Back* (DFB). The characteristics of these diodes differ, as shown by their characteristic curves in Figure 3-10.

Since the laser bandwidth is important in the calculation of the effects of dispersion in fiber-optic systems, it may not seem vital to discuss LED light sources where the solid-state laser diode is the optical transmitter of choice for cable television systems. Many buildings, however, are already wired

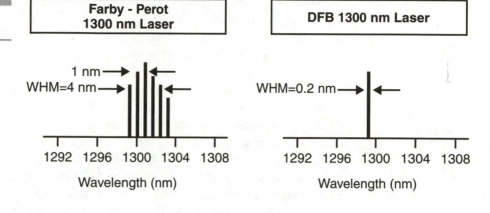

Figure 3-10
Spectrum of lasers

WHM is Width at 1/2 Maximum

using multimode fiber. In many instances, apartment buildings as well as commercial office buildings contain unused optical fiber. Such fiber could be used to connect services to subscribers since many of the distances are short compared to ordinary cable systems' fiber-optic plants. Small multimode optical detectors could feed the RF signal spectrum to a mini-distribution coax system or directly into a converter.

Light Emitting Diodes   Solid-state light-emitting diodes were used in many early data communication optical fiber systems. Much was learned about these light sources and many improvements were made. The LEDs had several problems associated with their use as an optical transmitter. The first was speed. The LED cannot be modulated at rates (speed) as high as a laser. Also, LEDs do not have as high an optical output as do lasers.

Since speed and intensity are important factors for light sources acting as transmitters, it should be evident why lasers are the best choice. Generally, light-emitting diodes have a much wider spectral width than lasers; this increases the effects of modal dispersion in multimode systems. Dispersion causes the stretching of short pulsed signals in time, which severely limits the pulse or data rate if it becomes excessive. Light-emitting diodes are made to operate in the 850 nm and 1,300 nm regions, which are in the lower operating loss regions for glass optical fiber.

Lasers   The solid-state laser diodes used in most single-mode long-distance fiber-optic systems can be made to operate in all the low loss optical windows of most fibers. The 1,300- and 1,500-nm regions are used in cable television single-mode fiber-optic systems. Both LEDs and laser diodes can be directly modulated by controlling the drive current. Both the LED and the laser have operating characteristics that vary with temperature. LEDs in general can operate within the range of -25° to +125° C. Lasers are somewhat more temperature-sensitive. Therefore, cooling using a Peltier cooler may be required in some applications. Usually when dealing with laser diodes in cable television systems operating in the usual head-end air-conditioned environment, proper heat sinking is adequate. Typical operating curve of optical output versus a drive current is shown in Figure 3-11.

An examination of this figure indicates that the operating region is the linear portion of the curves for both the LED and the laser. Notice that the temperature characteristics are different and the LED is operational over a wide temperature range. The extreme linearity of the laser diode and the high optical power output are important factors that make the laser a superior transmitting device.

Laser diode light sources are available as a single device or as a packaged unit mounted in a heat-sink enclosure. Most laser diodes in cable television applications appear in system optical transmitters that include the associated stabilization and drive circuits. The RF signal consisting of the band or bands of cable television channels is connected to the input. The optical transmitter converts this signal to a modulated light signal that is coupled to the fiber-optic transmission cable. At the transmitter output, the optical signal can be divided by an optical coupler into several optical fibers feeding different parts of the system.

Figure 3-11
LED and laser
characteristic curves

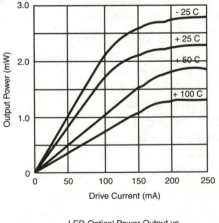

LED Optical Power Output vs.
Forward Drive Current
(Typical)

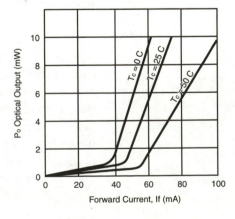

LASER Optical Power vs.
Forward Current
(Typical)

Optical Detectors

Equally important is that the received optical signal at the receive point be accurately detected and connected to the cable distribution system. A variety of photodiodes with a multitude of characteristics have to be considered in making a choice. One of the more important characteristics for a photodiode detector is its responsiveness. This in effect is the sensitivity for the photodiode. Responsiveness is given in amperes per watt. The watt figure is the optical power falling on the photodiode's junction, which produces a current through the photodiode.

Another parameter that is important is the rise time. The rise time is the time it takes for the current to reach its stable value as a sudden increase in light falls on the diode junction. Photodiodes are reverse-biased so the electrical signal is essentially the current through the junction, causing the diode's resistance to change. When no light falls on the photosensitive diode junction, the resulting current is called the dark current.

Early Optical Detectors Optical detectors have gone through many improvements. Early detectors were essentially photocells that would produce a small voltage that would vary with changes in light levels. Some people might recall the cadmium sulfide photocell that varied its resistance with changes in light levels. The silicon solar cell produces an electrical voltage when light rays fall on the cell's photosensitive area. Most of the photo- or light-sensitive devices previously discussed either require significant amounts of light or are too slow to be of much value in fiber-optic communication applications.

Photodiodes Solid-state photodiodes are used as the light detector or receiver in fiber-optic communication systems. The photodiode's p.n. junction is reverse-biased, which removes current carriers out of the depletion region that blocks the current until light enters that frees the electron's hole pairs, causing the current to flow through the junction. Doping can change the dark current of the photodiode and thus low light level operation can be improved. Since the transistor consists of either an n.p.n. or p.n.p. junction, if one of the junctions, usually the base-emitter junction, becomes the photo diode, the photo transistor results. Furthermore, if the photo transistor is placed with a second transistor into a Darlington circuit, the Darlington photo detector results.

As might be suspected, all of this transistor gain results in an increase in responsiveness. Again, it might seem that this type of light detector would indeed be superior, but the junction capacitance causes the rise time to increase. This, of course, affects the high-speed pulse (data) response and high-frequency response. Photo transistor detectors are ideal for high sensitivity and medium- to low-speed applications.

Pin Diodes   The pin-photodiode does not have the responsivness of the so-called amplified photo detectors, but it has a very small rise time and can be used for high-speed, high-frequency applications. The pin-photodiode is usually the one used for single-mode operations in cable television optical receivers. Usually, an optical receiver circuit uses an operational amplifier following the pin-photodiode, which does not load the photodiode output and provides some power gain as well as a low impedance output to the following circuitry. Companies that manufacture optical transmitters and receivers for cable television applications usually offer optical transmitters with input levels and a number of cable channels specified. Optical receivers are usually paired with the optical transmitter.

The cable operator now has to design for a given power budget calculated by the system losses, the transmitter output, and the receiver threshold sensitivity. An example of an optical transmitter and receiver fundamental circuit is shown in Figure 3-12. A listing of various photodiode detector characteristics is shown in Table 3-1.

Splicing and Connectorizing

Splicing and connectorizing optical fiber into a network is crucial to the operation of a fiber-optic communication system. Since the optical fiber exhibits such low loss per kilometer or per mile, a system should not allow losses to accumulate from the necessary splices and connectors. Prudent designers should order sufficient optical cable for segments that require a branching or termination point. Since connectors are required at stations in the system, extra splices along the way should be avoided. The quality of splicing and of connectors should be good so that the accompanying loss is held to a minimum. A variety of connectors are available for connecting a system, but a good design dictates that the variety of connectors should be minimized. This means that personnel will not have to be trained for different connectors. Also stocking a variety of connectors and the necessary tooling can be avoided.

Mechanical Splicing   One of the first splices to be used was a mechanical splice. Essentially, the optical fiber was cut, cleaned, and cleaved so the end was cut flush (perpendicular to the axis). When both ends were thusly

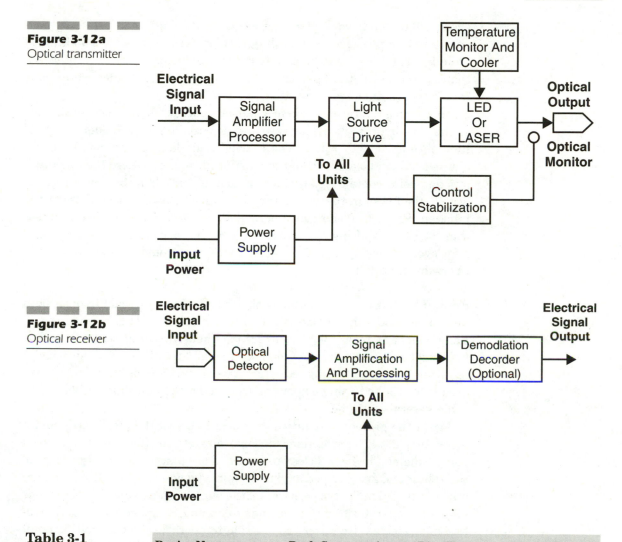

Figure 3-12a

Optical transmitter

Figure 3-12b

Optical receiver

Table 3-1

Photodiode detector characteristics

Device Name	Dark Current nA	Rise Time	Responsiveness
(Si) pin-photodiode	10.0	0.1-5ns	0.5 A/W
(In Ga As) pin photodiode	0.1-3	5ps-5ns	0.75 A/W
(Ge)Avalanche photodiode	400	0.25-1ns	0.6 A/W
(In Ga As) Avalanche photodiode	30	0.3ns	0.75 A/W

prepared, the ends were placed in a ferrule (tubing) and butted together end to end. Improvements in such types of splices included the addition of an index-matching gel to help mate the cleaned ends and the addition of a cement to prevent the fiber end from pulling away. Ferrules were made from either metal alloy or plastic.

Another big improvement in mechanical splices was what became known as the elastomeric splice. This type of splice used a plastic holder that held the splice open with a tool. The cleaved and cleaned ends of the fibers entered from opposite ends of the plastic splice piece where they met at a junction point containing an index-matching gel. When the tool was withdrawn, the fibers in the grooved track clamped closed to the fibers at the junction. Properly made elastomeric splices often yielded splice loss less than 1 dB. The splicing of optical fiber was always considered to be permanent even if such splices of the elastomeric type could be opened and the fibers disconnected.

Fusion Splicing   The method of choice for splicing optical fibers for both multimode and single mode is called fusion-splicing. This method essentially melts and fuses the glass fibers together. The equipment required to fusion-splice optical fibers is a sophisticated and relatively expensive device requiring proper operating expertise. Fusion-splicers, as they are known, are made to splice a single fiber at a time or, for the new flat or ribbon fiber cable, several at a time.

Again, the procedure requires the fibers to be cut the prescribed length, have any plastic covering removed right down to the glass, and be cleaved using a diamond wheeled cleaver. The ends are placed on the work stage of the splicer and are clamped in place by spring-loaded clamps. The fibers are essentially laid in "V" type grooves on the work stage. An optical microscope allows the operator to move the ends to an end-to-end connection. Activation of an electric arc raises the junction temperature to the melting point and then the arc is turned off. The junction of the two fibers becomes fused.

Present-day splicers place a sharp beam of light into one end of the splice joint while a photo detector at the other side detects light passing through the splice. This test allows the operator to measure the light loss through the splice. If the splice loss measured is too much, the operator can then redo the splice.

Fiber-optic splices for single-mode fibers are placed in a protective tubing that contains a metal strength member. This protective tubing is the heat-shrink variety and many fusion-splices have a heated drawer so this tubing can be shrunk down immediately after being spliced and tested. Spliced fibers in their respective tubing are clamped in splice trays containing extra loose fiber placed in a figure-eight channel.

Testing of the splice loss is completed right after the fusion has cooled and then a stress test is performed so the operator knows that fusion actually took place and the fibers weren't just placed together. This type of test is done by the splicing device. A slight tension is placed on the splice to test if indeed the fibers were fused. Following this test, the protective sleeve is slid over the fused junction and is heat-shrunk down.

The optical viewing method has been improved so most splicing equipment has a miniature television camera attached to the microscope and the image is projected on an LCD screen. Some splicing gear even has a video output that allows the operator the ability to connect a large screen monitor to the equipment. The monitor can be used in training other prospective splicers in the proper techniques. A rudimentary sketch of a fusion-splicer is shown in Figure 3-13. Also Figure 3-14 shows some examples of typical splicing termination equipment. The trays containing the splices have input and output ports for the fiber-optic cable to enter and exit the splice enclosure. The enclosures are usually the PVC plastic-sealed aerial kind or the plastic-sealed cabinet type for surface mounted pedestals.

Figure 3-13
Optical fiber
fusion splicer

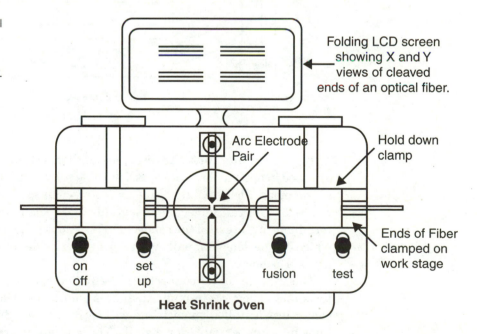

Folding LCD screen showing X and Y views of cleaved ends of an optical fiber.

Arc Electrode Pair

Hold down clamp

Ends of Fiber clamped on work stage

on off   set up   fusion   test

Heat Shrink Oven

Figure 3-14
Splice termination

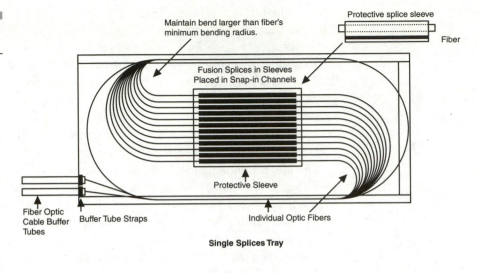

Maintain bend larger than fiber's
minimum bending radius.

Protective splice sleeve

Fiber

Fusion Splices in Sleeves
Placed in Snap-in Channels

Protective Sleeve

Fiber Optic
Cable Buffer
Tubes

Buffer Tube Straps

Individual Optic Fibers

Single Splices Tray

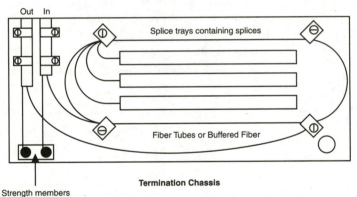

Out   In

Splice trays containing splices

Fiber Tubes or Buffered Fiber

Termination Chassis

Strength members

Types of Connectors   As discussed earlier, several types of connectors
are used in optical fiber networks. For cable television systems using mostly
single-mode fiber, mainly three types of connectors are used (see Figure 3-15). Optical connectors essentially allow the bare end of the fiber to
terminate at the end of the connector. This end is polished after installation,
which can be done manually with a polishing holder or mechanically using
a power polisher.

The installation of connectors, like fusion-splicing, requires sufficient
tools and expertise. In many instances, short pieces of optical fiber with
connectors installed on each end will be purchased. Such fiber pieces are
known as jumper cables or pigtails. These jumpers are usually cut in the
center and are fusion-spliced to the optical fiber. The optical fiber is then

Figure 3-15
Types of connectors
commonly used on
cable television fiber
optic equipment

FC

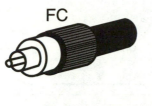

Typical insertion loss 0.5 - 1.0 dB
Push in & turn to lock connector.

D4

Typical insertion loss 0.2 - 0.5 dB
Screw type connector.

SC

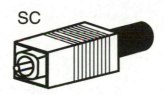

Typical insertion loss 0.2 - 0.5 dB
Push = pull type connector.

connectorized and is simply connected to the optical transmitter or
receiver as required. Such a procedure avoids having a connector crew to
connectorize a system, especially if not many connectors are needed. For
systems requiring an optical patch panel, it is best to have it professionally
connectorized.

In a typical case, fiber cables enter a termination station where the individual fibers in their color-coded buffer tubes are spliced to a single-fiber
cable with a connector installed. These single connectors usually run
through tracks or channels to the electronic equipment racks containing the
optical receivers and/or transmitters. If this station is simply a receiving station, the optical signals are connected to the receivers where optical to electrical (electronic) conversion takes place. If the station is a receive/transmit
station, then communication services can essentially undergo a drop/add
and be transmitted to other stations.

Pig-Tail Connectors and Terminations   Quite recently, the fiber-optic
coupler or splitter has been perfected. This device is actually an optical
fusion of more than just two optical fibers. Figure 3-16 illustrates the
concept of splitting and coupling the optical signal. Since this device is
made from optical fiber itself, it is quite small and physically compact. Such

Figure 3-16
Optical power
coupler

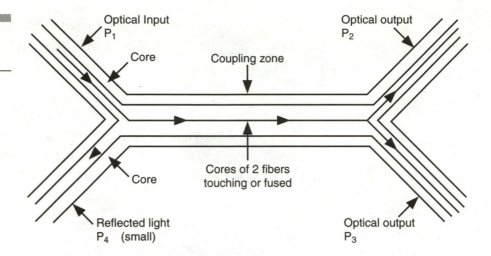

optical signal splitters can be placed in either aerial splice closures or in aerial optical receivers. They can also be found in terminating stations where an optical signal enters at a high level, drops off a low-level signal, and exits on another optical cable to another receiver station.

A device that is often needed is an optical attenuator used to reduce a signal into a receiver so as not to overdrive it. Optical attenuators can be made by making a calibrated bend in the optical fiber. Here light escapes from the fiber right through the cladding and plastic coating. Lost light results in an attenuated signal into the device connector. Such devices are made by buying plastic forms with a bending channel radius built in with the calibrated bend. The fiber is placed in the groove and the cover is closed. Another method of making an attenuator is with some fusion-splicers. Here the core of the joined fibers is offset an appropriate amount, causing a misalignment to decrease the signal the desired amount. Such splicing gear allows the operator to program in the amount of optical attenuation in dB.

Optical transmitters, receivers, splices, connectors, splitters, couplers, and attenuators are all the devices needed to put together an optical fiber communications network. Such networks and their terminations can be very large, employing various stations containing a great number of cables and components. Since miniaturization is often desired, optical stations can be mounted in surface cabinets and pedestals or in aerial-mounted housings and closures. Many design choices and methods are available to those charged with designing a system.

Fiber-Optic Trunking and Cable Television Applications

The first and obvious use of optical fiber in cable television applications was to replace the trunk cascade of coaxial cable amplifiers. This concept gave cable television systems some enormous advantages. The main advantage was the elimination of nearly all the system noise and distortion contributions due to the cascade of amplifiers. Secondly, eliminating the trunk meant lower power consumption with less related costs and signal leakage from the trunk system.

The only drawback was the fiber-optic signal was essentially light and was not a RF signal. At each node or many nodes, conversion from the optical signal to an RF signal had to take place. Some progressive urban systems using the subsplit reverse or mid/high-split upstream transmission replaced the downstream (forward) trunk with optical fiber and did away with their reverse system. As it turned out, few systems made this change and those that did pulled the forward amplifier modules and used the old trunk system for only the upstream application. The difficulty systems had facing this problem depended in large part on the availability of equipment types their manufacturer had available to rework the reverse-only system.

Systems considering optical fiber additions should study the methods of fiber-optic network topology as well as their cable system network topology. Fiber-optic technology has many similarities to coaxial cable systems. Both cables experience loss, which is a function of frequency, but fiber optic cables have a difference in loss versus frequency due to the material (glass) characteristics. Still, in systems operating at wavelengths of 1,310 nm or 1,550 nm, the loss is given in dB/km and the calculation of signal levels is done in much the same way as coaxial cable. For optical systems, the signal levels are in optical dB/km. The usual unit is in dBm of the optical power level. Hand-held optical power meters are used to measure system levels at cable ends.

Fiber-Optic Cable Overlay

The method of a fiber-optic cable overlay of the coaxial trunk system is the simplest approach in implementing fiber-optic techniques. This method is shown in Figure 3-17. It displays the optical fiber following the trunk route with often two or more fibers serving an area.

Fiber-optic cable is manufactured with many choices of single fibers placed in buffer tubes. For aerial applications, loose-tube cable is often the choice. Usually, there may be two to eight fibers per buffer tube and as

Figure 3-17
Fiber backbone
example

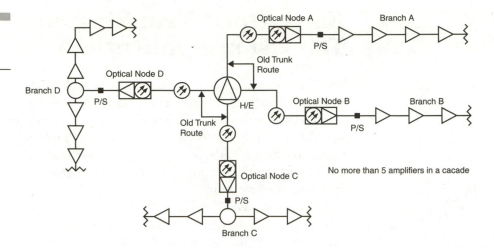

Figure 3-17
Fiber backbone
example

many as eight buffer tubes per cable. This can result with as many as 64 fibers in a cable. For the fiber trunk method, also known as fiber backbone, 24 total fibers are usually adequate with plenty of spares (dark fibers) for the future expansion of either up- or downstream requirements.

For cable systems carrying 50 or more channels, some terminal equipment will split the signal band into groups of channels and will optically transmit, for example, 16 television channels per fiber. The reason is, if the whole channel lineup is transmitted on one laser transmitter on one fiber, the laser power is spread over the whole band. When a laser transmitter transmits 16 6-MHz channels over one fiber, the power is spread over about a 100-MHz band ($6 \times 16 = 96$). This results in a higher receive power per channel. The network branches, as shown in Figure 3-17, indicate the optical paths that may actually consist of one or more fibers, each carrying a group of television channels. These groups of channels will be recombined at the optical receivers. It should be apparent that the addition of node connecting fibers will result in redundant optical paths to each node. Such alternate optical signal routing is shown in Figure 3-18. Signal source switching at each optical node can be controlled by a computer at the hub/head-end or by a simple loss-of-signal switch.

Development of Nodes and Sub-hubs   The development of system node locations is important in restructuring the network for system upgrades and increasing signal reliability. Alternate fiber-optic routing can result in each node's capability to maintain a reliable, nearly fail-safe system. It is imperative that each node be powered by a stand-by or *uninterruptible power sys-*

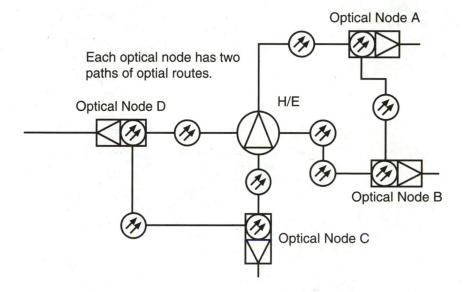

Figure 3-18
Alternate routes
connecting nodes

Each optical node has two
paths of optial routes.

Optical Node A

Optical Node D

H/E

Optical Node B

Optical Node C

tem (UPS). Failure of individual coaxial cable distribution amplifiers will only cause a small number of subscribers to lose the service. When the calls come in reporting an outage at a node, often the area and the device location can be determined quickly. Thus, the outage can be corrected in short order. System distribution nodes are another name for a sub-hub where signals from the main source or head-end are supplied by an optical fiber system. The optical-to-electrical (RF) conversion takes place at the node or sub-hub. Signals are then distributed on a normal coaxial cable system to the subscribers' homes.

Carrying this concept still further in the fiber-optic mode results in what is known in the industry as *Fiber to the Curb* (FTTC). The optical signal is converted to RF signals within the pole- or pedestal-mounted tap. The subscriber's home is connected to the tap port via a coaxial drop cable (see Figure 3-19). At present, few if any cable television systems operate in this manner. The day will no doubt come when the fiber optical signals will be dropped to the subscribers' home and the optical-to-electrical conversion will take place within the set-top modem. This concept results in what is known as *Fiber to the Home* (FTTH) and is shown in Figure 3-20.

Fiber-optic Return Methodology   Since optical fiber has essentially replaced the trunk coaxial cable cascade, some means is needed to supply the return or upstream signal requirements. Instead of using the old trunk

Figure 3-19
Optical fiber to RF
taps (FTTC)

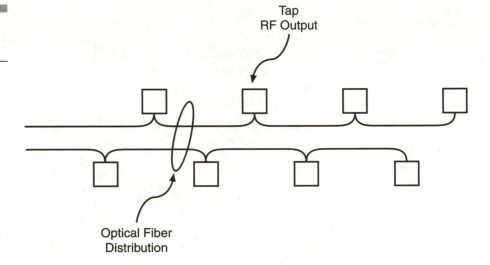

Figure 3-19
Optical fiber to RF
taps (FTTC)

Figure 3-20
Optical signal
distribution to
subscriber terminal

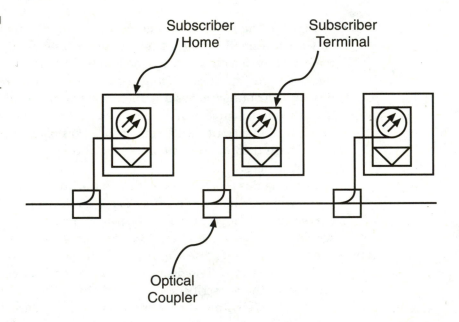

system for the return path, a better solution would be to use an optical fiber system. Enough extra fibers should be included in the cable to cover the return signal needs.

Basically, the upstream system is often called the reverse-tree concept where signals originate from the subscribers and are collected along the path to the central hub/head-end where they in turn are sent or redistributed to their destinations. Normally, most subscriber signals consist of ordering information for pay-per-view service or possibly utility meter-reading information or alarm information. This type of use does not require high bandwidth or high data-rate signals. If telephone service is being contemplated or if computer interconnecting will become a requirement, more upstream supporting bandwidth will be needed. From a practical standpoint, the upstream service most likely should be digital. Presently, telephone service as offered by the Telco operators functions in the digital domain beginning with the basic T-1 system. Signals from subscriber locations can be collected via RF, usually the 5- to 40-MHz subband is converted to digital signals at each sub-hub or node and sent to the main hub/head-end on optical fiber. Each subscriber will have a unique access code contained in the digital header that will enable the hub/head-end to recognize who is sending or receiving messages. Other header data in digital form will contain routing and control information.

Several brands of head-end and distribution equipment that address the reverse digital system are presently available. As in the past, several of the well-known companies have product and application information as well as customer engineering support available to cable operators considering such plant expansions and upgrades. Also, several consulting and design companies have suitable software for design and mapping for both fiber and coax systems. Cable systems using both fiber and coax methods are known as *hybrid fiber coax* (HFC) systems.

It is imperative that cable television systems considering staying in the business a long time must add fiber-optic technology to their systems in order to compete in the telecommunications enterprises during the next century. At this writing, AT&T, the largest telephone service company, and *Telecommunication Inc*. (TCI) are merging, thus pooling their interests in providing *Internet Protocol* (IP) services including telephone service.

The upstream reverse system can be a simple straightforward node to hub/head-end or a more complicated node to a sub-hub and then to the main hub point. Many manufacturers and cable operators have and/or offer several elaborate upstream systems. Figure 3-21 illustrates the node-to-hub system concept. If the return system consists of digitally modulated optical carriers, the digital circuitry at the hub/head-end must decode the digital data stream and assemble the data packets to be sent to the desired destination. If the return data is collected on subband RF carriers, the whole subband can be transmitted on an optical return carrier. This may be simpler than demodulating the return carriers, assembling a high-speed return signal, and transmitting back to the hub/head-end as a digital

Figure 3-21
Node to HUB/HE
concept

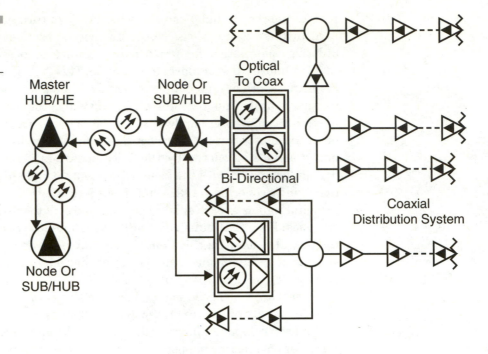

optical signal. Cable systems considering such system enhancements should do a comprehensive survey of available return system components from the various manufacturers.

Fiber Optical MDU Systems   Many cable television systems in urban or suburban areas are fortunate to have large apartment or condominium complexes, known as multiple dwelling units or simply MDUs. These large complexes offer high counts of equivalent dwellings per mile, which is important criteria for system plants. The internal or inter-building wiring is, in essence, an extension of the distribution system and can add several amplifiers to the cascade. When fiber-optic technology is being introduced into existing coaxial cable systems by replacing a trunk, extending the optical fiber into the complex is advisable. Thus, a node or mini-hub point fed by optical fiber should be situated within the building or a group of buildings depending on the existing coaxial cable network topology.

For high-rise apartment complexes, the fiber node could be located in some location at the middle-floor level. The optical signal can then be converted to RF television carriers and be fed to distribution points on the other floors. In some cases, an optical coupler can be connected to a receiver on the first floor and another receiver on the middle floor. The receiver on the first floor can distribute signal from the first to the middle floor and the receiver at the middle can distribute signals to the top floor. Thus, many design choices are available with various trade-offs. Often, for each situation, the best choice is the obvious one. Figures 3-22 and 3-23 illustrate a couple of design examples.

Since many apartment or condominium complexes have affluent professional people as residents, high-speed computer data service will be welcome. This means that the two-way return system has to be active. Therefore, the fiber-optic return system should be installed to assemble subscriber return signals and connect then to the hub/head-end for further routing. The downstream system with its large bandwidth will send the digital information back to the requesting subscribers. This high-speed connection to the Internet will surely be attractive to many subscribers. It must be remembered that the high-bandwidth cable system drop is far greater than the copper twisted pair of the phone drop.

Fiber-Optic Super Trunking

Optical fiber systems are ideal to use in supertrunk situations. Supertrunking methodology has been used by many cable operators as a good solution to a variety of problems. Such applications consist of satellite-receiving connections to the hub/head-end point or of connecting hubs together. Connecting hubs together improves the signal reliability but adds significant power and maintenance costs. Also, more coaxial plants mean more possible points for RF leakage and ingress problems. Replacing supertrunk systems with an optical fiber system provides all the good points and few of the bad ones.

Head-end to Head-end Trunking and Earth Station Feeds The satellite to hub/head-end connection is the most common use for a supertrunk. It should be realized that a supertrunk system is just a signal transportation method from one point to another, usually in one direction. Only one fiber is typically required to carry signals from a satellite station to a hub, but most manufacturers' cables contain a number of fibers in color-coded buffer tubes. When optical fiber is planned to connect the downstream signal and upstream return path from the mini-hubs or nodes, a single fiber or a series of them can be used to pick up the signal from the satellite-receive station to the master hub/head-end. This concept is shown in Figure 3-24.

Systems using a coaxial cable supertrunk connecting a satellite station to a hub/head-end a distance of five miles could use an amplifier cascade of two or three amplifiers, depending on the cable size. Such a cascade should

Figure 3-22
Optical to RF coaxial
on 1st floor

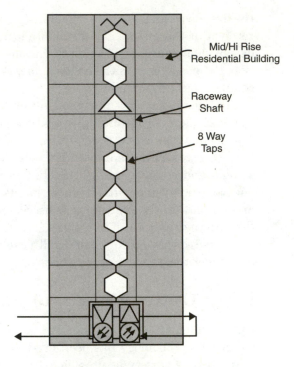

Mid/Hi Rise
Residential Building

Raceway
Shaft

8 Way
Taps

Figure 3-23
Optical to RF coaxial
mid-level

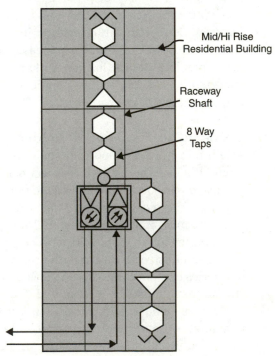

Mid/Hi Rise
Residential Building

Raceway
Shaft

8 Way
Taps

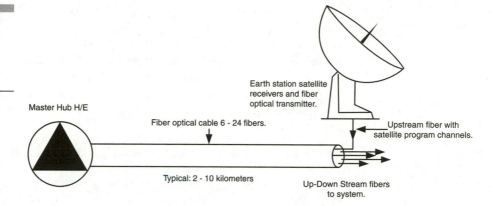

Figure 3-24
Fiber optic earth receive station to master Hub/H/E system

Earth station satellite receivers and fiber optical transmitter.

Master Hub H/E

Fiber optical cable 6 - 24 fibers.

Upstream fiber with satellite program channels.

Typical: 2 - 10 kilometers

Up-Down Stream fibers to system.

have its own power supply so that it is not dependent on the downstream path for its power. The frequency plan for transmission from the satellite station can be lowered to the standard VHF band for the upstream run to the hub/head-end in order to be in the lower-loss end of the coaxial cable.

At the hub/head-end, the satellite channels have to be converted for the required downstream channels, yet all this conversion adds noise and distortion to the signal. When fiber-optic methodology is used, the "on-channel" assignments can be made at the satellite station and transmitted to the hub/head-end. The optical receiver will supply the "on-channel" RF signal to the combining network for downstream distribution. From this example, the benefits of using optical fiber technology should be evident.

Television Broadcasters to Head-end   As cable operators start to utilize fiber-optic technology, more and more uses become evident. When installing fiber-optic cable to replace the coaxial trunk and expand the upstream return bandwidth, extra fibers are often included in the cable for future use. In many instances, these extra fibers (dark fibers) may not be seen as necessary at the time of installation, but their use will become evident later on.

One such use is to connect to the local broadcast television stations by optical fiber. One should recall that early cable systems' main reason for operating was to improve the reception of broadcast television stations. Using optical fiber to receive television broadcast stations from distances up to 50 kilometers is indeed possible. By doing so, the quality of the programming will certainly be high. Gone are the industrial sparklies, weak signal snow, and lightning interference. Even if the station's transmitter fails, the cable signal should still be there. Employment of the *emergency broadcast service* (EAS) would also be enhanced since all stations are required to carry this service and soon all cable systems will also be required to participate.

The system design for broadcast station pickup is nearly the same as for the satellite receive station pickup, shown in Figure 3-21. Variations of this can be configured, depending on the location of the studio and/or transmission facility. Preferably, the station would have available a video and audio feedpoint so the cable operator would have a choice of transmitting baseband video and audio or use an on-channel television modulator transmitting optically to the hub/head-end location.

Benefits of Fiber-Optic Plant Addition

By now, it should be evident that there are indeed great benefits to cable television operators who install fiber-optic systems. The benefits include cost savings, signal quality improvements, and signal reliability. Cable television operators have been striving for years to improve the product and control costs, both of which are attainable by adding optical fiber technology. In this section, we'll take a brief look at these factors that are involved in adding fiber optics to cable systems.

Cost Comparisons   Most cable operators have problems controlling costs when making plant improvements that have to be balanced between the costs and the prospective earnings and/or savings. Since one single-mode fiber in a multifiber cable has the capacity to carry more information bandwidth than a coaxial cable, a cost comparison is very difficult. Also, a single fiber connecting services over normal cable system distances requires no repeater amplifiers or electronics between the transmitter and receiver. Thus, comparing the amplifier coaxial cable cost per mile with a single fiber might be a more meaningful comparison. However, in aerial and underground systems, a normal fiber-optic cable contains several fibers placed in several buffer tubes. Hence, a single- or dual-fiber cable is not normally manufactured for aerial or underground plants.

For those more mathematically inclined, a more analytical approach would consist of a calculation of cost per hertz of the bandwidth. Such a procedure might be useful in convincing management to invest in an optical fiber system upgrade. A rudimentary cost analysis for a run of cable starting at the hub/head-end that travels five miles into the system may give some cost comparison between coaxial and fiber optics. Excluding labor costs for strand and cable placement is a reasonable consideration since it is assumed to be the same for coaxial and optical plants. For 750-MHz cable service, the cost factors and calculations are shown in Example 3-1 for a

Cost comparison of
coaxial system vs.
fiber optical system
for 5 miles of cable
plant

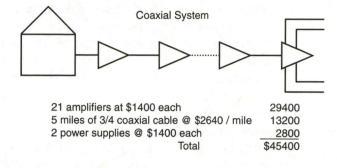

Coaxial System

21 amplifiers at $1400 each	29400
5 miles of 3/4 coaxial cable @ $2640 / mile	13200
2 power supplies @ $1400 each	2800
Total	$45400

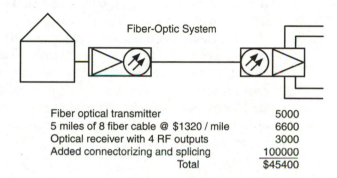

Fiber-Optic System

Fiber optical transmitter	5000
5 miles of 8 fiber cable @ $1320 / mile	6600
Optical receiver with 4 RF outputs	3000
Added connectorizing and splicing	100000
Total	$45400

five-mile sample run. It must be remembered that the design topologies
between the two methods can become quite different.

Signal Improvements   One of the benefits of fiber-optic systems is the
improved signal quality. The technical staff of most cable television systems
are well aware of the noise build-up and distortions of a cascade of amplifiers. As the upper frequency limit of coaxial cable systems increases, the
number of amplifiers per mile increases. This fact in turn limits the length
of a cable run or system reach. Some operating cable systems that employ
long cascades of costly and power-hungry feed-forward amplifiers may find
it impossible to extend the bandwidth of such systems without doing a
fiber-optic rebuild.

   Some system operators use a microwave feed from a master head-end
facility and beam programming to several sub-hubs serving the areas with
cable service. Each sub-hub has a microwave receiver and electronic equipment that converts the microwave signal to the normal cable television
channels. The master head-end may have several transmitters and antennas to beam the signal in several directions. Such a system avoids the

amplifier cascade distances and provides improved signals to the sub-hub distribution points. Maintenance of such systems, along with following the FCC rules of compliance for the facilities, is usually costly. Some cable systems have replaced the microwave facilities with optical fiber connecting the sub-hubs to the master head-end.

Picture Quality Improvements   One of the most important benefits of adding optical fiber in place of coaxial system trunk is that system reliability is improved. As stated earlier, often no electronic devices are used in most fiber-optic runs in distances of up to 30 miles. Thus, there is nothing that can fail or cause a problems, unless it is subject to a disaster such as a fire or having a pole hit or the cable cut.

Some cable operators serve a node or sub-hub by dual optical paths, thus adding a redundant feed to each distribution point. If problems do occur, it usually is the coaxial plant that causes the outage. Optical fiber runs do not contain connectors that can pull out of amplifiers or line-passive devices that cause outages. Also, RF leakage or ingress does not happen in fiber optical systems. Therefore, signal quality and reliability are maintained to superior standards.

Fiber-Optic Construction and Installations

Fiber-optic cable installation methods have a lot in common with coaxial cable installation, but they have differences as well. One must remember that the information-carrying optical fiber is glass, which is brittle and sharp and can shatter, while coaxial cable is metallic. Both types of cable employ plastic jackets for protection from water and handling. Coaxial cable has mechanical strength in its center conductor, outer aluminum sheath, and its plastic jacket. Fiber optical cable has either a metallic or tough plastic strength member and outer jacket that adds strength. As mentioned earlier, the optical fibers are contained in color-coded buffer tubes either tightly or loosely and are appropriately named loose-tube or tight-tube construction. In most cases, loose-tube optical cable is used in aerial plants, while tight-tube cable is often used in inter-building or underground plant applications.

Handling Fiber-Optic Cable

Handling optical cable carefully is important in order to assure no damage will result and the cable will have a long service life. Optical cable should not be bent beyond the minimum specified bending radius. It should also not be subjected to heavy objects being placed on it or running over it. Such action can stress or damage the glass fibers within the cable. Also, the cable should not be subjected to physical shocks either from hitting the cable or dropping a reel on the pavement. Most of the damage to a fiber-optic cable occurs during the loading-unloading delivery phase or during installation.

Loading and Unloading   The initial loading of cable is done at the manufacturing plant and is usually done correctly using proven methods and equipment. If the shipment to a cable operator is direct from the manufacturer on the manufacturer's truck, minimal problems can be expected. Fiber-optic cable that is shipped to a distributor and then to the end user usually has to go through two or more load-unload cycles where more possible damage to the cable may result. Therefore, the cable system operator should be aware of where and how the cable is being shipped.

Most cable manufacturers and sales materials list instructions for loading-unloading procedures in their catalogs. Generally, all manufacturers of optical fiber cable state that cable reels should stand on the edge of the reel and be rolled in a specified direction. The reels themselves are made of either steel or heavy wood. A steel reel can usually be returned to the manufacturer for credit, while the wooden reels may not be returnable. The wooden reels are fully lagged and strapped with two steel bands around the lagging. The lagging consists of wooden slats nailed to wooden reel flanges. This protects the cable from being smashed, as shown in Figure 3-25.

Unloading from a trailer truck can be done with a ramp, which can be tricky because the reel of cable has to be restrained from rolling on the ramp. Probably the best way is with a crane or a fork-lift truck. Placing a pipe through the reel hole and using a crane or fork-lift to lower it to the ground works well. This is usually done a reel at a time. Usually a reel of optical cable contains more footage than coaxial cable; hence, the number of splices are held to a minimum. Some manufacturers' trucks are equipped with either a crane or a lift-tailgate, making the unloading procedure simpler.

Figure 3-25
Example of a properly
made reel of optical
fiber cable

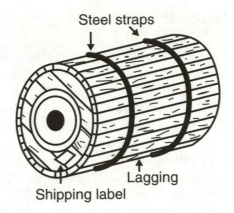

Figure 3-25
Example of a properly
made reel of optical
fiber cable

Fiber-Optic Cabling Handling   Proper storing of optical cable is important and necessary while the optical system is being installed. Fiber-optic cable should be left standing on the reel edge or rolling edge and should never be stacked reel-on-reel on its side. The fiber-optic cable should be separated from coaxial cable so it is not mixed up by the installation contractors and handled unnecessarily.

Indoor storage is best for optical cable since wet and cold weather can cause deterioration of the wooden reels. Systems storing optical fiber in a warehouse should use either an appropriate loading dock or a forklift truck to place the optical cable reels on a reel trailer for installation.

Admittance Testing

As with most commercial companies, testing incoming material should be performed and any equipment or material ordered that does not meet the necessary specifications can be returned. This process prevents installing faulty equipment that produces a poor system or product. Admittance testing of materials should be performed on all equipment, not just cable. The testing of optical cable for loss (attenuation) and effective length can be made quite easily and cable manufacturers make both ends of the cable accessible on a reel. Therefore, testing the cable on a reel-to-reel basis can be accomplished.

Methods of Testing   Testing for cable attenuation can be made simply with a power meter and light source. Some sources and meters can have the bare ends of the fibers cleaned and cleaved. Since this process is done on each fiber, placing connectors on each one is a big job. Some cable operators test one or two fibers in a cable and assume if they test well, then the rest must be OK. It is best to test each fiber individually. This procedure is shown in Figure 3-26. Both ends of the cable have to be accessible for this method. The main advantage is the low cost of the optical source and power meter.

A better method of testing for optical attenuation is to use an *Optical Time Domain Reflectometer* (OTDR). This method shows any abnormalities in attenuation for the whole length of the optical fiber. Since OTDR instruments cannot make measurements close to the instrument cable connector, a length of a single fiber is placed between the instrument and the reel of fiber-optic cable. Essentially, this moves the cable reel to be tested away from the instrument by the length of the single fiber, which is usually approximately one kilometer in length. This so-called blind spot of the OTDR is referred to as the dead zone. This single-fiber cable is tested and its results are documented before testing the cable reels. This procedure is shown in Figure 3-27.

Measurements and Records Once the testing of incoming fiber-optic cable has been completed, the results should be recorded and filed in an appropriate manner. This information could prove useful if a problem

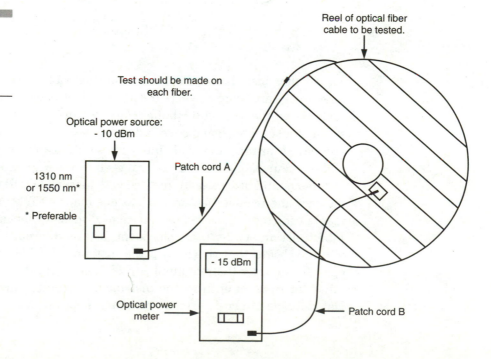

Figure 3-26
Fiber optic earth receive station to master Hub/H/E system

Figure 3-27
OTDR testing of fiber-optic cable

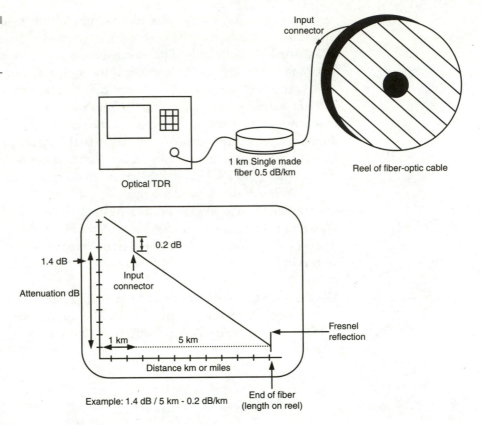

Example: 1.4 dB / 5 km - 0.2 dB/km

occurs after installation. The documentation can simply be a list of each fiber and its attenuation where the fiber is identified by the color of its outer coating and the color of the buffer tube containing it. Each reel also has a serial number that should be recorded. Some cable operators develop a form and have it printed and made into pads so they can be numbered, torn off, and filed. An example of such a form is shown in Figure 3-28.

A faster and more elegant method is to have an OTDR that has a digital memory and read-out feature such as one made by Tektronix. Such instruments can store measured data in digital memory for later examination. Fiber color and buffer tube color identification information can be entered via a keyboard that comes with the instrument. This process allows the operator to enter the reel number and measured attenuation, along with other information such as the instrument, operator, and time and date. Data storage can be done on a convenient floppy disk.

████ ████ ███ ███
Figure 3-28

Example of a manual
record of test data for
fiber-optic cable

Company Name, Address, and Telephone Number

Test date: _____   Name of Technician: _____

Cable reel no: _____   Order no: _____   Date of delivery: _____

Type of installation: _____   Inst. model no: _____   Calib. date: _____

Test wavelength: _____ nm

Addmittance Test Data

Buffer tube Color	Fiber Color	Optical Att. dB	Buffer tube Color	Fiber Color	Optical Att. dB

Fiber-Optic Cable Installation

Installing fiber-optic cable requires some special procedures that are different than the procedures used for installing solid aluminum-sheathed coaxial cable. Most people remember quite well that coaxial cable should not be crushed or kinked during handling and that the installation process can cause the cable to be stressed more than the loading-unloading process. Therefore, the differences in handling will be addressed for both the aerial and underground installation process.

General Procedures   In general, fiber-optic cable is manufactured in longer lengths than coaxial cable for two reasons. First, it is not necessary to place amplifying or repeating devices along the cable run, and secondly, it is most important to minimize the number of splices. Splicing fiber-optic cable can be an extremely long and laborious process since each fiber has to be individually fusion-spliced and placed in a splice tray. The splice trays are then placed in cylindrical PVC weather- and moisture-proof enclosures.

The long lengths of fiber-optic cable have to be installed in sections. Therefore, pulling tension on the cable has to be closely monitored in order to not exceed the recommended value by the manufacturers. A process called figure-eighting is often used, which helps relieve pulling tension. This process is shown in Figure 3-29.

Contractors who are more experienced and specialize in optical fiber cable installation have equipment that controls the tension on the cable. Such equipment is a servo-controlled winch or capstan drive where the cable tension controls the action. Pulling grips used for optical cable are the same type as used for coaxial cable and distribute the tensile pull load over the outer cable jacket.

Since splicing is to be avoided, service loops (excess cable) are placed along the cable run. The initial loop is placed at the beginning of the run and often appears approximately every fifth or sixth pole span. This excess cable can be used when the aerial pole plant is rerouted when a new road or road-widening work is needed. Thus, cutting and splicing in a piece is avoided. This excess length has to be designed into the system and initially taken into account when calculating cable attenuation and optical input/output levels. These service loops use a turning frame at each end of the loop that resembles a snowshoe or tennis racquet. Essentially, a figure-eight pattern of cable is stored in the loop and is raised and mounted on the steel aerial strand. An example of a service loop is shown in Figure 3-30.

In general, fiber-optic cable should not be tightly lashed with coaxial cable so precautions should be taken when pulling the lasher. It is vital to find a contractor experienced in proper fiber-optic cable installation.

Figure 3-29
Figure-eight tension relief method

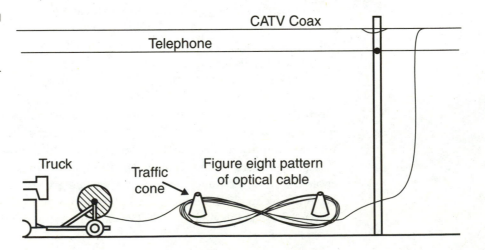

Figure 3-30
Excess cable stored
in service loop

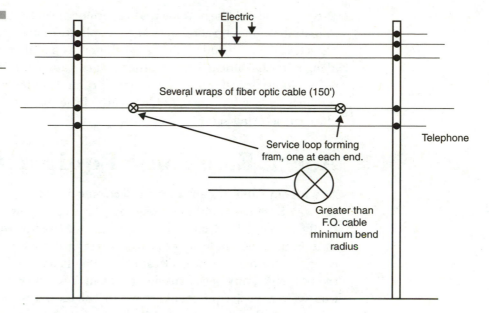

Figure 3-30
Excess cable stored
in service loop

Installation Equipment and Procedures Underground cable installations seem to be used more often in today's urban and suburban environments. Underground cable installations use a variety of methods and techniques from direct burial in a trench to a conduit system. Conduit appears in a rigid pipe form of PVC or semi-flexible corrugated plastic tubing. Some conduit manufacturers can supply the cable preinstalled in the flexible conduit, which can then be buried in a pre-made trench or plowed directly.

In city or urban areas, optical cable is pulled through existing ducts. In such installations, a pulling tape is placed in the duct by a compressed air method. Next, the fiber-optic cable is connected to a cable grip, which in turn is fastened to the pulling tape. A winch at the other end pulls the tape through the duct followed by the fiber-optic cable. A pulling lubricant is applied at the source end of the cable run that aids in relieving the pulling tension. The pulling tension should be monitored at the pulling end by a dynamometer or by a servo control on the winch. The maximum pulling tension should not exceed the value recommended by the manufacturer. The splice locations for the underground plant should be placed conveniently for servicing. Some of the equipment used was shown earlier in Figure 1-9 in Chapter 1.

Again, it should be obvious that experienced contractors should be hired to install an underground optical cable plant. In areas where trenched or plowed cable is installed, it is a good idea to install a metallic conductor along with the optical cable in order to facilitate identifying the cable path with a locating instrument. If a coaxial cable is to be installed along with the fiber-optic cable, the metallic sheath of the coaxial cable can be used to place the locating signal.

Aerial Electronic Equipment

As with any cable system, certain electronic devices have to be placed along the way. Such electronic equipment requires some power source and it certainly cannot get it from a glass fiber. Either the fiber optical cable has to contain metallic conductors to carry power or it must be obtained from the coaxial cable plant. Since optical fiber cable has replaced the coaxial cable trunk system, any power needs are taken from the coaxial cable distribution system. Such equipment consists of the receivers that convert the optical signal to the RF signal for distribution to subscribers. At these locations, the reverse signals are assembled and are either converted to digital signals or are transmitted in the analog domain in the sub-band through an upstream transmitter using upstream optical fibers.

Types of Electronic Equipment

At locations where the fiber-optic cable feeds signals to the coaxial distribution system, the optical signal may terminate or be passed through to a location further downstream. An optical coupler can be used to divide the signal to feed the optical receiver with an appropriate optical level and pass a larger level on further. This is essentially the same method as the RF directional coupler. Such a method is described in Figure 3-31. An optical transmitter can be incorporated in the terminal package to retransmit the optical signal further downstream.

Splitting and Retransmitting Several manufacturers make optical couplers that essentially allow a tap leg and a through leg. Such devices are usually fusion-spliced into the system and are placed within the housing containing the optical receiver. Since most aerial housings are two-pieced with a hinged cover and a weather-sealed gasket, as well as a R.F.I. gasket, the optical portion usually occupies one side and the RF the opposite side. This is similar to housings used in coaxial cable systems with the usual RF

Figure 3-31
Split downstream
signal from optical
coupler. Upstream on
seperate fiber

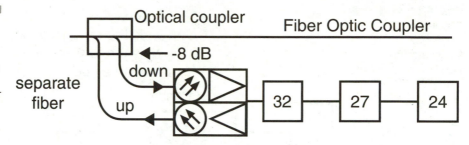

Figure 3-32
Nodes being fed
cable service on
seperate fibers.
Nodes can be strand
mounted (only
forward system
shown)

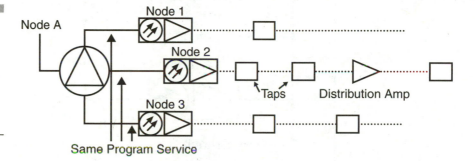

test point available for signal-level measurement. The advantage of placing pigtails and connectors to the output ports of the optical coupler permits the measurement of the optical signal using a power meter. In many cases, the optical signal provides service on several fibers connecting to several distribution nodes, as shown in Figure 3-32. Nodes can be sub-head ends or hubs where locally generated programming can be inserted. Some systems distribute the signals unscrambled to the hub sites where local signals are inserted and premium channels are scrambled.

Optical to RF Receivers   The optical receiver placed in a node or sub-hub location converts the optical signal to an RF signal consisting of all the television carrier frequencies. A nominal RF output level is usually between +30 and +35 dBmV, with most manufacturers at +32 dBmV. Often the maximum and minimum optical signal levels that provide a range of RF output signal levels are specified. Equipment manufacturers usually have a variety of options available to the system designers. Such options may be a single, dual, or quadruple RF output. Thus, the optical receiver system can act as an RF bridging amplifier, providing a signal level up to four outputs. Another feature is the return option that assembles the upstream RF carriers and converts them to an optical signal for transmission to the head-end hub via

Figure 3-33
Single fiber
forward/return
signals

Branch A:
λ_1 is the forward optical signal
λ_2 is the forward optical signal on
the same fiber.

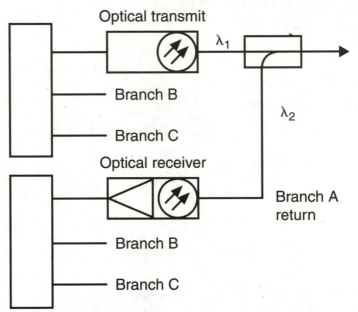

Optical transmit

λ_1

Branch B

Branch C

Optical receiver

λ_2

Branch A
return

Branch B

Branch C

λ_1 is the forward system optical carrier wavelength.
λ_2 is the return system optical carrier wavelength.

an optical fiber. If the upstream signal operates on a different wavelength than the downstream signal, the same fiber can be used to carry both up and down signals. In this case, a splitter-coupler combination has to be placed in the fiber at the head-end/hub and at the node location. Such a system is illustrated in Figure 3-33.

System Powering Considerations The optical equipment placed at node or sub-hub locations usually receives electrical power from the coaxial cable RF plant. This power source is the usual 60- or 90-volt a.c. power from the normal cable plant. Since this optical receiver is critical for providing service to the node distribution point, a stand-by or UPS power supply is necessary.

It must be remembered that system reliability is a requirement for any system operating in today's telecommunications environment. In many instances, the sub-hub or node is not a strand-mounted or pedestal-mounted unit but appears in a building housing local signals. Dropping services and adding services are performed at these locations and since commercial power is available the optical receivers are rack-mounted where the RF channels are used to provide selected programming, as required by the sub-hub. The sub-hub location should have a stand-by power source to provide emergency power. Often at these locations, RF television signals provide service to the immediate area, while remote service nodes are served by optical fiber.

An early approach to emergency powering of remote electronic equipment was to use a rechargeable battery pack built into each device. This plan did not work out because most available battery packs did not provide enough on-time to keep the device operating during prolonged outages. Possibly the day may arrive when battery packs are improved and power requirements are reduced, making this method practical.

It should be evident that all service vehicles should have a substitute commercial power source installed. This makes it possible for vehicles to provide commercial substitute power to the stand-by power supplies when the batteries are exhausted.

Optical Equipment Tests

Remote optical equipment appears either as strand- or pedestal-mounted devices servicing sub-hubs or mini-hubs. It is here where the conversion from optical to RF television carriers take place. Some systems house their terminal equipment in buildings where commercial power is available. Such equipment is usually rack-mounted. At these locations, optical and RF signal levels can be deleted or added.

Optical Measurements   Measurement of optical signals at the receive sites is necessary to establish the optical signal input to the receiver. When setting up a system, the transmitter at the head-end/hub is turned on and the optical signal is measured at its output. When the optical fiber is connected to the transmitter, the optical signal level is measured at the receive point. The difference in levels is the cable loss. These levels are measured with an optical power meter indicating the level in dBm of optical power. The difference in dBm readings from transmitter to receiver is the optical cable loss in dB.

Next, the signals are connected to the transmitter and the power level in dBm at the receiver is recorded. This value can be written on a label at the receiver for future reference. Now the RF signals can be measured and adjusted, as required by the coaxial cable system. Such adjustments are for the level and slope of the television carriers. Most optical power meters are battery-operated and operate in either a single- or dual-wave length mode. Such optical power meters have a variety of features such as the input connector type, a set of adapters, and extra battery packs, as well as a carrying case. Some power meters have an accompanying light source available to aid in making fiber-optic cable-loss tests. As stated earlier, read-out is in optical power level in dBm (one mw reference). These instruments cost anywhere from $500 to $1500, depending on the model and options.

Coaxial Cable RF Measurements   Once the proper optical levels have been established and compared to the predicted design parameters, the RF system has to be adjusted. As most of us know, these measurements can be made using a signal-level meter. For this method, measurements are made on a channel-by-channel basis. Some signal-level meters have a wideband tuning feature that presents all the carrier levels on a wide LCD display similar to a spectrum analyzer. The method of choice would be to use a high-quality spectrum analyzer, preferably one with digital storage. Measurements can be made and recorded on a printer either on location or later at the office. These printouts are a useful maintenance tool for monitoring system performance. Long-term data can be stored on floppy disk for comparing signal levels over time. A high-quality spectrum analyzer, as used by many cable systems, can measure signal-to-noise ratios as well as signal distortions. Cable operators using fiber-optic technology usually can see the difference in signal quality at the node sub-hub locations as compared to the cascade of amplifiers previously connected to these points.

Digital Technology and Cable System Applications

A Short History of Digital Communications

Most of us feel that it is becoming a digital world with all the computers, cellular telephones, and the forthcoming digital television. Actually, digital technology has been with us a long time. Digital technology involves the binary numbering system with two digits, or two states, such as On/Off. Connecting and disconnecting circuits is the business of the telephone companies and it should come as no surprise that much of the development of digital technology came from the telephone companies, mainly the well-known Bell Labs.

Two-state communication was used by primitive tribes in the forms of smoke puffs or drum beats. Morse code was a form of digital code where long tones and short tones were used in combination to determine alphanumeric characters. A human operator either wrote down the message or typed it on a typewriter. Morse code was used by ships at sea using radio transmitters and receivers to communicate with shore stations or other ships. Long-distance communications were realized using such techniques quite reliably. Marine radio used morse code for ship safety almost to this date. Recently, morse code marine radio has been terminated and ships at sea no longer require a qualified operator on board. Now high-speed satellite communication systems operating with digitally coded message packets provide marine communications.

Nature of Digital Technology

Since digital systems operate in a two-state manner, digital communications are similar to a conversation in which one can ask only yes or no questions. Because two states can be described by two voltage levels, such as 0 volts and +1 volt corresponding to a binary 0 and binary 1, then a train of pulses can describe a sequence of binary numbers. It is these pulse sequences that describe digitally encoded alphanumeric characters. A pulse train of 0-volt (binary zero) and +1-volt (binary one) pulse sequences transmitted through a wire will have a d.c. level that is positive and varies with time. Therefore, reworking the digital pulse stream to a better form will save transmitting power levels through a wire or cable. Two types of digital sources communicating through a communications channel are shown in Figure 4-1. In Figure 4-1a, the communications channel is digital, and in Figure 4-1b, the communications channel is analog. These types essentially are the methods used for computer communications through the telephone system.

Figure 4-1

Digital
communication
examples

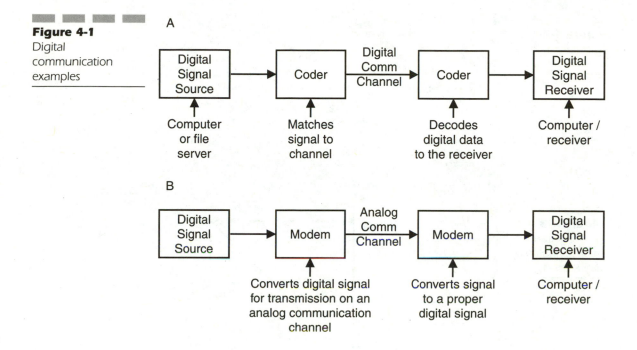

A

| Digital Signal Source | → | Coder | Digital Comm Channel → | Coder | → | Digital Signal Receiver |

Computer or file server

Matches signal to channel

Decodes digital data to the receiver

Computer / receiver

B

| Digital Signal Source | → | Modem | Analog Comm Channel → | Modem | → | Digital Signal Receiver |

Converts digital signal for transmission on an analog communication channel

Converts signal to a proper digital signal

Computer / receiver

Binary Numbers & Decimal Numbers As mentioned earlier, digital systems are based on a two-symbol binary numbering system in which sequences of pulses of voltage levels correspond to binary digits. In this system, a direct relationship exists between a decimal magnitude and a binary magnitude. Thus, one can convert from one to the other, as shown in Figure 4-2. It is this relationship between the decimal and binary numbering systems that allows the analog value to be converted to a digital (binary) value. Analog-to-digital converters basically take an analog time-varying electrical signal, sample it, and then convert each slice or sample to a digital number representing the amplitude of the sample. Figure 4-3 describes this process.

It should be noted that each sample value is held from the sample signal to the next sample. During this time, the ramp signal is still increasing, which results in an error in the voltage value between samples. Increasing the sampling rate, thus decreasing the sample time, will result in less errors. Notice there is a four-bit digital number value for each sample, so doubling the sample rate will double the number of samples and the samples per second.

Figure 4-2

Decimal to binary conversion

For the decimal quantity fo 24 the digital number corresponding to 24 is computed as follows using the scale shown below:

Least significant bit

Most significant bit→ 7   6   5   4   3   2   1   0   Bit number

2^7 2^6 2^5 2^4 2^3 2^2 2^1 2^0 Value

128 64 32 16 8 4 2 1

If all 8 bits (0-7) are used to describe decimal 24:

16 + 8 = 24 = 0001100

Figure 4-3

A voltage ramp signal in both analog & digital form

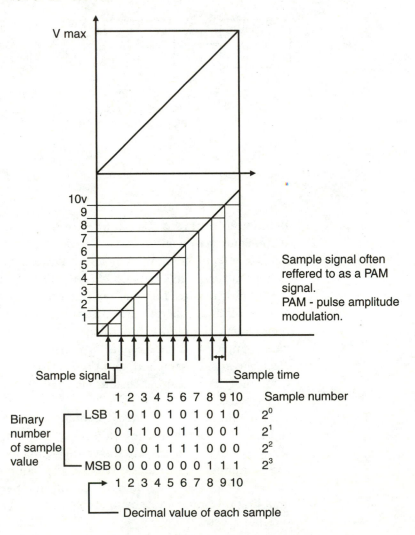

Sample signal often reffered to as a PAM signal.
PAM - pulse amplitude modulation.

Converting analog voltage values to digital numbers is a common technique used in digital voltmeters and multimeters. The digital numbers are then displayed as decimal digital numbers. A chip set is used in the much used digital multimeters.

Electrical Representation of Binary Systems   Now that it has been shown that electrical values of a voltage or current can be represented in the digital domain as a sequence of digital numbers, many uses and methods can be developed. Each digital number value can be represented by two voltage levels corresponding to a digital one or zero. Using the example of Figure 4-3, the sequence of the 10 samples can be represented, as shown in Figure 4-4. This digital pulsed signal has a positive one volt as a binary one and a zero volt value as a binary zero. This signal is often referred to as unipolar (either 0 or 1V) and so has a varying d.c. value, depending on the amount of binary ones (positive pulses). This can become a problem when transmitting digital information through a communication system.

Another technique that reduces the d.c. level is to use a +1V for a binary one and a -1V for a binary zero. This method tends to reduce the d.c. level value and is often referred to as bipolar. These pulse trains can also split the signal time of a bit and return to a zero value before changing to the negative value representing a binary zero. This type of signal is referred to as RZ (or return to zero). Several pulsed signals representing the same binary

Figure 4-4
Pulse digital signal

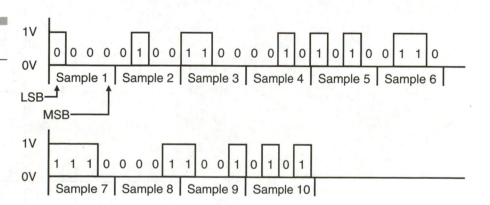

A 1 value = 1V
A 0 value = 0V

number are shown in Figure 4-5. Circuits using such pulsed signal wave-forms maintaining bit synchronization are most important. Clearly, the form shown as *Bipolar Return to Zero* (BPRZ) gives more signal transitions, thus aiding bit (clock) synchronization. These forms of pulsed signals are known as line code formats.

Coding and Decoding   Sequences of digital pulses correspond to binary digits, which in turn represent, for instance, alphanumeric characters that

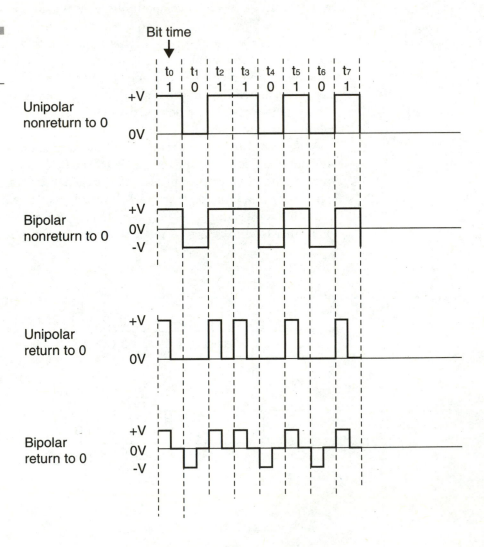

Figure 4-5

Some examples of line coding format

require another coding/decoding procedure. Certain applications require a number of bits to form what is termed a computer word or a data byte. A seven-bit character sequence is used to form the *American Standard Code for Information Interchange* (ASCII) code commonly used in computer-to-printer operations. This code is shown in Figure 4-6. Table 4-1 lists the abbreviations and their definitions.

Figure 4-6
ASCII code bit/
character chart

				b_7	0	0	0	0	1	1	1	1
				b_6	0	0	1	1	0	0	1	1
				b_5	0	1	0	1	0	1	0	1
Bit Number					Column							
b_4	b_3	b_2	b_1	ROW	0	1	2	3	4	5	6	7
0	0	0	0	0	NUL	DLE	SP	0	@	P	'	p
0	0	0	1	1	SOH	DC1	!	1	A	O	a	q
0	0	1	0	2	STX	DC2	"	2	B	R	b	r
0	0	1	1	3	ETX	DC3	#	3	C	S	c	s
0	1	0	0	4	EOT	DC4	$	4	D	T	d	t
0	1	0	1	5	ENQ	NAK	%	5	E	U	e	u
0	1	1	0	6	ACK	SYN	&	6	F	V	f	v
0	1	1	1	7	BEL	ETB	'	7	G	W	g	w
1	0	0	0	8	BS	CAN	(	8	H	X	h	x
1	0	0	1	9	HT	EM	)	9	I	Y	i	y
1	0	1	0	10	LF	SS	*	:	J	Z	j	z
1	0	1	1	11	VT	ESC	+	;	K	[	k	{
1	1	0	0	12	FF	FS	,	<	L	\	l		
1	1	0	1	13	CR	GS	-	=	M	]	m	}
1	1	1	0	14	SO	RS	.	>	N	^	n	~
1	1	1	1	15	SI	US	/	?	O	_	o	DEL

SP - SPACE   DEL - DELETE

Table 4-1

Definitions of ASCII abbreviations

Column 0 Abbreviation	Definition	Column 1 Abbreviation	Definition
NUL	All characters zero	DLE	Data link escape
SOH	Start of heading	DC1	Device control
STX	Start of text	DC2	Device control
ETX	End of text	DC3	Device control
EOT	End of transmission	DC4	Device control
ENQ	Enquiry	NAK	Negative acknowledge
ACK	Acknowledge	SYN	Synchronous idle
BEL	Attention alarm	ETB	End/transmission block
BS	Backspace	CAN	Cancel
HT	Horizontal tab	EM	End of medium
LF	Line feed	SS	Start/special sequence
VT	Vertical tabulation	ESC	Escape
FF	Form feed	FS	File separator
CR	Carriage return	GS	Group separator
SO	Shift out	RS	Record separator
SI	Shift in	US	Unit separator

Books have been written on various codes that contain many features, yet such features are not a concern to communications people as long as the type of line code is compatible with the transmission medium. Certain types of codes contain error detection and/or correction features, but the ASCII code is widely used in personal computer input/output operations and has undergone several revisions since it was first adopted in 1963.

Other alphanumeric codes are the old *Teletype* (TTY) code and the one developed by IBM called the *Extended Binary-Coded Decimal Interchange Code* (EBCDIC), which is an eight-bit code with many characters.

Error-checking is performed on ASCII code using another bit, making a character seven bits plus one parity bit for a total of eight. Bits 0 through six are the character's seven bits and bit seven is the parity bit. The EBCDIC code uses no parity bit. Error-checking techniques are described in several of the references given in the bibliography at the end of the book. Several chip manufacturers make chip sets used for encoding and decoding both ASCII and EBCDIC codes.

Digital Methods for Data & Graphics   Digital methods used for handling computer data and graphics are commonly used today in a variety of applications. The computer monitor has to display the ASCII-coded letters and numbers on a cathode-ray tube screen in its proper place and color. The monitor's digital circuitry has to contain chip circuitry that corresponds to a grid system of columns and rows used to position the digital coded data. Since the ASCII data is used for an alphanumeric character, the computer and monitor program will place it on the next character position on the screen. This same program will give a screen background color along with pre-programmed prompting messages. Again, the digital chip makers have produced the necessary circuitry used to produce screen graphics.

This same screen technique is used to display digitally encoded television pictures. The specification given by some manufacturers for resolution is in *dots per inch* (DPI), which is an older printer type specification for resolution. A more current specification is given in pixels with 2048×2048 being the present specification for computer color monitors. Comparatively, this is far better than the resolution of most present high-quality television sets.

Digital data values are often plotted on graphs or curves by digital plotters that require data handling by a computer. Under software control, the computer addresses the plotter with instructional information as to scale, size, and so on and feeds the plotter with digital data that is in turn plotted by a moving pen or a dot matrix. Manipulating data for graphic presentation is big business in today's computer environment.

Digital Computing and Data Storage

A personal computer can be found in many homes in America and the Internet, along with what is called E-Commerce, is becoming commonplace. Computer manufacturers are producing PCs working at faster speeds and with huge amounts of memory. Large files can be stored on hard disks with humongous capacity. Programs can be purchased on a CD, thus not requiring space on a hard drive. Most present-day computers found in many homes and offices use many forms of storage, such as CD-ROMs, diskettes, and solid-state *random access memory* (RAM) chips.

Modem units, either installed within the computer or as an external unit, allow the computer to be connected to the telephone (telco) system, which in turn connects to other online systems. It is here that the cable television system has more to offer in access speed than the telephone system. In brief, the coaxial cable drop has more high-speed data capacity than the copper-twisted telephone drop.

The Role Of The Microprocessor   The heart of most PCs is the main microprocessor *integrated circuit* (IC) chip that controls the data transfer and computational speed. At present, the fastest microprocessor operates at a clock speed of 450 MHz. Communications people, particularly those in the field of cable television, think of 450 MHz as a 60-television-channel service. The main circuit board containing the main microprocessor chip is referred to as the motherboard. This circuit board also contains supporting circuits to the main microprocessor. Most of the electrical signals at 450 MHz occur in and close to the microprocessor chip, thus limiting signal radiation to quite low levels. As the main microprocessor chip speeds increase, the input/output data rates increase as well. This requires more and more frequency bandwidth from the communications provider.

Transferring Data & Mass Storage   Transferring data from one work-station to another in a rapid manner is a requirement of many enterprises. Large data files loaded and unloaded between user computers and mass storage facilities are commonly used by businesses. Downloading programs consisting of large data files is done often. Many cable television systems operating today have the capacity to deliver high-speed data traffic to users connected to the system. Thus, the transfer of large quantities of data can take place a lot faster by cable operators with available channel capacity.

In the past, typical cable operators avoided commercial areas simply because they believed no subscribers were present and thus decided not to place their system there. Cable operators interested in expanding their businesses by offering data communication services to commercial enterprises will have to extend their plants to these areas. In short, go where the business opportunities exist.

At present, many banks, department store chains, and gas stations, to name a few, use VSAT systems to provide data communication services. Most of these systems, however, operate in one direction with the upstream through the *local exchange carrier* (LEX). Cable systems should be able to provide such services to such customers.

Development of Data Communications

Providing high-speed data communication services to commercial businesses should provide much added revenue to cable system operators willing to upgrade and expand their distribution systems. Many cable systems already have been exposed to digital signal transmission from the use of addressable converters. Data communications using digital signal transmission will be used with many of the new cable modems for HDTV and pay-per-view program selections. The progressive cable operator should install optical fiber in both the downstream and upstream path along with stand-by power supplies needed to increase system reliability.

Early System Computer Interconnects   Soon after personal computers appeared as office workstations, the need to interconnect computer systems together became apparent. The first method was an obvious solution of using the *Dual Tone Multifrequency* (DTMF). This method resulted in the acoustic coupler modem in which the telephone handset was placed in a cradle that would translate digitally encoded data into DTMF tones. Transmission could take place in both directions from the mouthpiece and the earpiece. Of course, data communications were slow.

Next, the separate modem system was developed using *Frequency Shift Keying* (FSK). The modem converted the digital data pulses into two frequencies representing ones and zeros. FSK data received by the telco network was converted to pulsed signals and sent to the receiving computer via its input/output port. Figure 4-7 illustrates modem development.

At present, most modems are plug-in cards inserted into the motherboard and connected directly to the telephone system via a modular telephone connector plug. As modems were required to transfer data at higher bit rates, the complexity of the modulation schemes increased. From simple FSK modulation with a data rate of 300 to 600 bits per second development led to binary phase shift keying, which increased the effective bit rate to 1,200 *bits per second* (bps). *Quadrature amplitude modulation* (QAM) increased the telco modem effective bit rate to the present standards. Current modem advertisements state the modem bit rate at 56 Kbps, qualified by the statement that the speed at 56 Kbps depends on the line quality of the *local exchange carrier* (LEX). Since a cable television system could easily provide data rates equal or greater than this speed, the development of a cable modem is necessary for a cable operator to enter this phase of service.

Figure 4-7
Development of
telephone-computer
modem

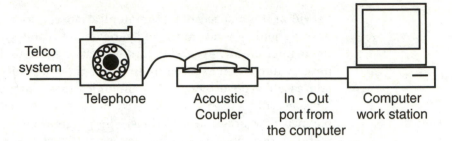

1. Early

Telco
system

Telephone

Acoustic
Coupler

In - Out
port from
the computer

Computer
work station

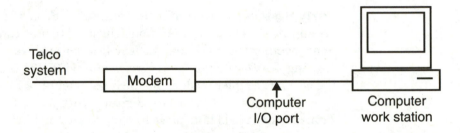

1. Later

Telco
system

Modem

Computer
I/O port

Computer
work station

The Local Area Networks   The use of telco modems allowed residential
and business computer systems to interconnect. A network of simple dial-
up modems and/or leased line systems produced an area network among
connected users. In-house networks where computers were connected via
buses, switchers and channel routers formed *local area networks* (LANs).

Present Data Communication Systems

Present-day data networks consist of networks of owned inter-company networks connected to national and worldwide telephone systems that form a
commercial telecommunication network. LANs are connected to *wide area
networks* (WAN) to various interconnected providers, including the telephone systems and other common carrier operators. It is here that progressive cable operators can provide a necessary and useful service to the
telecommunications industry.

It should also be noted that the principal professional organization representing the cable television industry changed its name to the Society of

Cable Telecommunication Engineers. This was indeed an appropriate action. Cable systems should be and are aligning their operations to be not only a provider of television programming, but also a computer-interconnection service to the digital world. Cable television operators in the not-too distant future will be an *Internet provider* (IP) made possible by a cable modem. Since the Internet is offering voice service, cable operators acting as IPs can also provide a form of telephone service.

LANs and Types Topology

LANs were developed as a necessity for supplying data connections required by the commercial business world. Since no standards were specified, technicians and engineers made rudimentary networks to solve particular problems. Network topologies were finally developed and several types of LANs were developed.

These network topologies exist in a variety of forms, such as a star network, where all end-user stations have to pass through a main switching center. Other topologies are the tree-branch network, the ring type network, and the bus network. These topologies are shown in Figure 4-8.

Figure 4-8
Network topologies

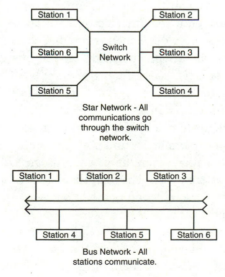

Star Network - All communications go through the switch network.

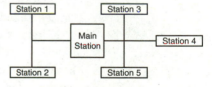

Tree Network - Stations on same branch communicate if not they go through the main station.

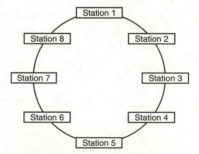

Bus Network - All stations communicate.

Ring Network Topology - Stations act as repeaters for each other.

Various techniques are also employed to control the network. The switched star network is controlled by the central switcher. For the tree network, each station has its address and control from branch to branch through the main station. A bus network is controlled by a collision detection system used to prevent all stations trying to communicate at once. The ring topology usually uses a token-passing scheme in which the station that needs to transmit must possess the token.

Each network topology has its advantages and disadvantages. The specific application often determines the one that will be best suited. A brief study of some of the more popular LAN systems follows in the next sections.

Ethernet   Ethernet is probably one of the oldest LAN types still in existence today and it seems as if it keeps reinventing itself. Basically, Ethernet is a bus topology and is one of a family of IEEE 802 architectures. Ethernet uses coaxial cable that supports data rates of 10 Mbps. This cable is a standard 50-ohms characteristic impedance. Ethernet rates over coaxial cable have been pushed to 100 Mbps, while for shorter distances over 10 Mbps copper twisted-pair can be used.

Recently, as the need for faster transfers of large data files has become critical, Ethernet systems converted to optical fiber. Thus, for distances of 200 meters, multimode fiber systems are adequate, while single-mode operation distances of approximately five kilometers are attainable at speeds up to 1 Gbps. The conversion to fiber was quite simple, following the same routes and often not requiring any electronics between stations.

Ethernet was originally developed through a concerted effort by Xerox, Intel, and DEC. Workstations operating on the bus cable had either an Ethernet modem or a *Network Interface Card* (NIC) installed into the workstation computer. Ethernet operates on the bus using *Carrier Sense Multiple Access with Collision Detection* (CSMA/CD). A station wishing to transmit has in the modem or NIC a circuit that looks for a data carrier on the bus. Seeing none, it transmits and if at that same moment another station transmits, the collision circuit in each of the transmitting stations' NIC will call for a time-out. The length of the time-out period varies between users. Once a connection is made between stations and a carrier is present, it prevents other stations from trying to transmit.

For many commercial businesses, Ethernet LANS provide the needed data transfers between computer workstations. Many Ethernet cable systems use a piercing pressure tap similar to early cable television systems that was used to connect the data transceivers to the bus cable line. An example of an Ethernet system is shown in Figure 4-9.

The Ethernet LAN is essentially a transmission line system and should be properly terminated. Branching has to be made with coupling networks

Figure 4-9
Ethernet bus system

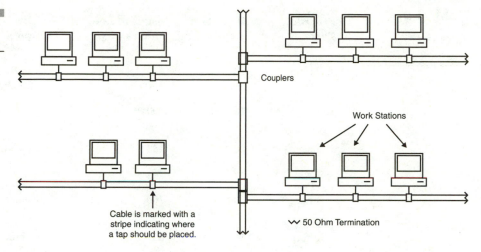

Couplers

Work Stations

Cable is marked with a
stripe indicating where
a tap should be placed.

〜〜 50 Ohm Termination

all properly impedance-matched. Repeater amplifiers can be placed in the line when needed to increase the carrier signal level. The data rate, or bit rate, is related to the data packet size by the transmission path length. For short distances of approximately 100 meters in length, 10 Mbps can be supported by an unshielded twisted dual-pair cable, which is actually telephone cable. This is very attractive for small inner-building LANs using low-cost, easily installed telephone wire. In many cases, the business occupants found excess telco wire already installed and in place. Thus, a low-cost 10 Base-T LAN could be quickly put together. Table 4-2 lists the generic progression of the Ethernet LANs.

The Ethernet software requires that data is sent in packets with the appropriate header containing addresses of the sender and receiver. The formation of a data packet is shown in Figure 4-10. The preamble section is often eight bytes of zeros and ones, providing enough transitions to establish bit synchronization (clock sync). Following the preamble is the destination address of six bytes. This address can be a specific station address or a broadcast for every station to receive or a selected group of stations to receive the enclosed message. The field type consists of two bytes describing the data field of 46 to 1500 bytes. The four remaining bytes are the cyclical redundancy-checking sequence.

The line code for Ethernet is what is called the Manchester code at the usual 10 Mbps. This code assures a pulse transition with every bit, thus maintaining clock synchronicity.

Token Ring   Another type of LAN is the Token Ring network, as shown in Figure 4-8. Each workstation is placed within the ring where the cable

Table 4-2

Ethernet LAN examples

Ethernet Type	Data Rate Mbps	Segment Length Meters	Cable Type	General Designation
10 Base 5	10	500	Coax RG-4 50 ohm	Thicknet
10 Base 2	10	200	Coax RG-58 50 ohm	Cheapnet
10 Base T	10	185	#24 4 wire Telco	
100 Fast Ethernet	100	100	50 ohm coax	
Gigabit Ethernet	1000	25	50 ohm coax	
1000 Base Lx	1000	220	MM1 fiber	
1000 Base Lx	1000	550	MM2 fiber	
1000 Base Cx	1000	25	Copper twisted-pair	
1000 Base Sx	1000	5000	SM3 fiber	

1 Multimode 62.5/125 mm fiber-optic cable

2 Multimode 50/125 mm fiber-optic cable

3 Singlemode 5/125 mm fiber-optic cable

Figure 4-10

Ethernet packet

Preamble 8 bytes	Destination address 6 bytes	Sender address 6 bytes	Field type 2 bytes	Data field 46 - 1500 bytes	CRC 4 bytes

passes through each station. When no station is transmitting, the ring rotates around from station to station. The token is actually a coded bit sequence. When a station is ready to transmit to another station, it captures the ring and then proceeds to transmit its data. Data is in the form of a packet of data bits. The transmitting station has to send data packets of specific sizes to fit the allowed transmitting timeslot, as determined by the LAN specification. The receiving station receives the token and returns it to the sender upon completion of the data packet transfer. The sending station then gives up the token to be circulated again, making it available to another sender.

Variations of the Token Ring LAN depend on what the equipment manufacturer provides. Naturally, most providers offer the usual bells and whistles, such as station priority levels and packet sizes.

Ethernet and Token Ring LANs have quite a few pros and cons. It should be obvious that substantial delays can take place, particularly for large rings since the token takes time to circulate. The ring delay time can cause a loss time equaling a large number of data bits. As the ring becomes loaded with data, a station can wait a long time before it can seize the token.

Token Ring LANs appeared around 1980 and gained acceptance rapidly since this type of network had less data congestion problems than Ethernet LANs. The IBM corporation introduced its version of Token Ring architecture around 1985 (IEEE Std 802.5). Today many Token Ring LANs are operating, providing adequate service required by users. The cable used in Token Ring networks is easily recognized as shielded twisted-copper pairs with a two-prong screw-on connector. Data rates of 4.0 Mbps to 10 Mbps are attainable on Token Ring LANs.

Broad Band Systems   Using cable television technology, a broadband LAN was developed. This type was referred to as *Manufacturers Automated Protocol-Technical Office Protocol* (MAP/TOP). This type was developed through the efforts of General Motors and various equipment providers. MAP/TOP was needed badly on the manufacturing floor for automobiles. Electrically, the production line is a huge electrical noise generator making twisted-pair and even coaxial cables or shielded-pair cable too full of noise to properly carry data. Since automatic or robotic manufacturing techniques require data communications with the controlling computer, some quiet communication methods were developed. MAP/TOP modulates high frequency data carriers with digital data where electrical noise is less, in the high frequency spectrum.

By stacking multiple carriers, as in cable television systems, large amounts of data could be handled by MAP/TOP. Using 75-ohm coaxial solid aluminum cable, the usual passive circuits such as taps, splitters, couplers, and line-repeating amplifiers provided the needs of such a design concept. Since data had to flow in both directions, a frequency band was assigned to the upstream direction and the downstream direction. An example is shown in Figure 4-11.

In this figure, the upstream band is 5 to 174 MHz for a bandwidth of 159 MHz and the downstream band is 234 to 400 MHz for a bandwidth of 168 MHz. This allows for nearly equal up-down bandwidth. As more data channels are needed for the downstream direction, the upper frequency limit can be increased to accommodate the need.

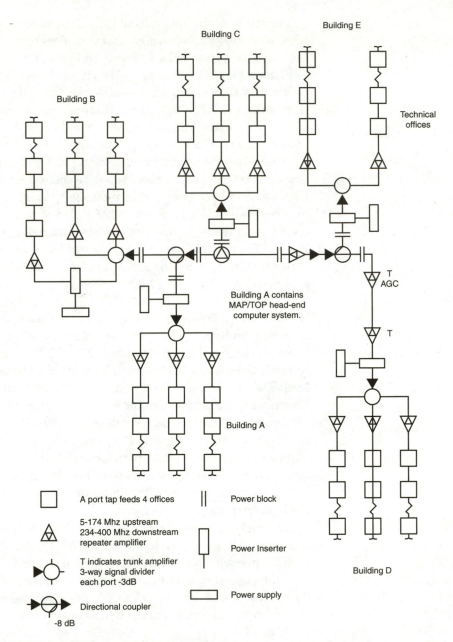

Figure 4-11
Example of a
broadband MAP/TOP
type local-area
network

Building C

Building E

Building B

Technical
offices

Building A contains
MAP/TOP head-end
computer system.

T
AGC

T

Building A

Building D

☐   A port tap feeds 4 offices

△   5-174 Mhz upstream
234-400 Mhz downstream
repeater amplifier

▶◯   T indicates trunk amplifier
3-way signal divider
each port -3dB

▶◯   Directional coupler
-8 dB

‖   Power block

▯   Power Inserter

▭   Power supply

Downstream signals containing modulated carriers that contain digitaly encoded manufactureing
instructions make up the forward signal. Messages are sent from the head-end to the
manufacturing machines and offices.

Upstream signals on modulated RF carriers contain data, requests for materials, and schedule
progress. These are sent through the head to other offices and warehouses.

The MAP/TOP LAN using coaxial cable television technology required a lot of maintenance including testing for signal leakage in the air safety frequency bands. Also, such a network consumed quite a lot of power and made keeping some of the equipment cool a priority. Many of these systems have been supplanted by fiber-optic networks.

Fiber-Optic Systems

Fiber-optic cable, as discussed earlier, has less optical signal attenuation than coaxial cable has for electrical television frequency attenuation. Thus, for most common physical cable distances, as found in commercial buildings and industrial parks, no electronic amplifier or signal repeaters are needed for a fiber-optic system. Therefore, most optical fiber systems are completely passive except for the terminating transmitter/receivers. As business computers sped up, the requirements for high-speed data transmission also increased and fiber-optic development provided the solution.

FDDI Fiber Optical Systems   One such system was the *Fiber Distributed Data Interface* (FDDI). The *American National Standards Institute* (ANSI) developed the standard X3T9.5 for the FDDI system that operates at 100 Mbps using a counter-rotating ring topology. As in most ring network topologies, a token-passing method is used to gain access. The FDDI network topology is shown in Figure 4-12.

It was necessary to develop a special connector for the FDDI network. Since both the primary and secondary fiber ring passed through each node, a dual fiber connector was developed, as shown in Figure 4-13.

An FDDI node operates as a repeater of data that is not addressed to that particular station. This technique allows for distances up to two kilometers between node stations. Many FDDI networks use a multimode-graded index fiber of the 62.5/125 millimeter variety, but this is not a hard and fast rule. Other fiber types can also be used. For most FDDI applications, LED light source transmitters and photo-pin diode receivers are adequate. For larger ring sizes with nodes further apart, single-mode fiber and laser transmitters are used. Often large corporate network LANs will use FDDI dual-ring networks as a backbone connecting other regional LANs. The interconnecting device is referred to as a bridge, which communicates between different network types.

SONET Systems   Another network that uses optical fiber technology is the *Synchronous Optical Network* (SONET), which was developed recently

Figure 4-12
Dual Ring FDDI
Network

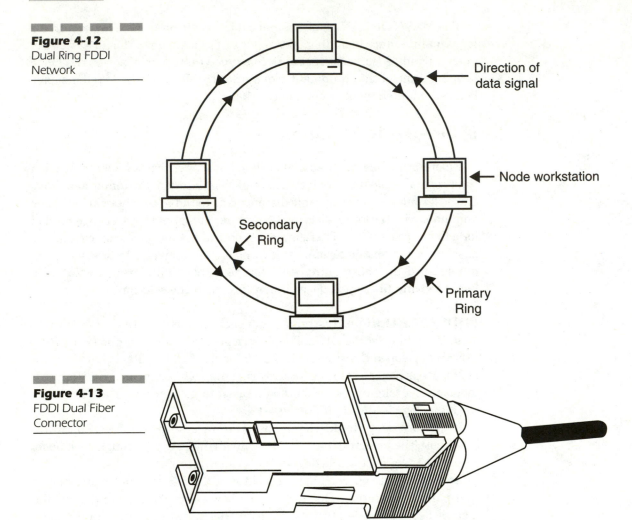

Direction of
data signal

Node workstation

Secondary
Ring

Primary
Ring

Figure 4-13
FDDI Dual Fiber
Connector

for the telecommunications industry. The SONET standard is for North American synchronous rates and a similar standard for Europe is the *Synchronous Digital Hierarchy* (SDH). SONET/SDH systems operate in the digital domain, are synchronous, and operate on single-mode fiber optical networks. The digitally encoded data is packet-switched and is organized into 810-byte frames including headers containing signal routing information. Data frames can be switched in and out individually without unbundling the SONET frame. The data transmitted using SONET technology can be digitally encoded telephone traffic and commercial data traffic, all intermixed on SONET frames. The standard specifies frame sizes and data rates and configures a ring topology that provides a redundant communication path. Thus, a SONET system is essentially self-healing, providing a very reliable network. Data rates for SONET systems are all multiples of each other, thus preserving the synchronization. Sonet systems will be revisited in other sections of this book as well as in Appendix C. Cable operators and their technical personnel should look at the possibilities of providing access to all these data carriers as new business opportunities and an accompanying source of revenue.

Telephone Systems and Digital Technology

 The largest and oldest telecommunication system in the U.S. and possibly Europe as well is the telephone system. Since its invention by Alexander Graham Bell, the telephone network has grown and progressed in size and technology continuously. The telephone system is known for its high quality and reliability, setting high standards for other industries to follow. The telephone industry, mainly *American Telephone and Telegraph* (AT&T,) produced many useful inventions and scientific discoveries through research by Bell Labs on speech, hearing, circuit switching, and electronic components. Such efforts produced valuable knowledge for the benefit of industry and the public in general.

Basic Telephone System

The basic telephone system consisted of copper pair cables installed on utility poles strung about the countryside, connecting the telephone sets in subscribers' homes to the central terminal location called the *local exchange* (LEX). This network remained essentially an electromechanical system for many years. When electronic amplifiers became practical and radio methodology was developed, the telephone companies applied these technologies to long-distance telephone service.

Presently, the telephone network is rich in fiber optics, which has essentially taken over the long-distance service and provided other services to commercial enterprises needing data communications. The cable television operator stayed out of commercial areas where the telephone systems were providing telephone communications. Oddly, the telephone companies expanded their fiber-optic plants into commercial areas and not into residential areas. The cable operator remains today with the residential subscriber coaxial cable drop with nearly a gigahertz of bandwidth, while the telephone company has its twisted-pair drop wire of limited bandwidth.

The Telephone Set The basic telephone set or instrument has grown through many stages of development. The rudimentary telephone network consisted of copper twisted-wire pairs insulated with silk, wax, and wax paper covered with a lead jacket. Local service was only provided initially. Each instrument or telephone set had a mouthpiece (transmitter) and an earpiece (receiver) with a switch hook to hold the earpiece. All calls initiated by lifting the switch hook connected to the manually operated local exchange office where a live operator requested the called party's number. A plug on the switch board was placed into the jack corresponding to the called party's number and the operator would turn on the ringing current.

Eventually, the operator and manual switchboard was replaced by a dial system that automatically switched the calling station to the called station's circuit and performed the ring, connect, and disconnect functions. The rotary dial was a serial pulse generator that operated the electromechanical (relays) circuit switch at the LEX. Long-distance or toll service to other LEXs was accomplished by trunk lines operated by toll switches. At first, these toll switches were manually operated by a switchboard operator who made the connect and disconnect, keeping track of time and charges. Later the toll service was operated by a switching system similar to that used at the LEX.

These early telephone systems were troublesome, mainly the outside cable plant. Breaks and cracks in the lead jacket caused water and moisture to enter and soak the fabric insulation causing cross-talk and false ringing. Specially trained cable splicers who could work with lead and hot solder were needed to service and extend the cable plant. The arrival of plastic insulated and jacketed cable essentially solved the lead cable problems.

Early telephone sets were electromechanical in design and construction. The "hybrid," as it was known, was a series of transformer phases wired to prevent transmitted voice currents from blasting the earpiece and incoming received voice currents from being dissipated in the transmitter. Telephone personnel referred to the hybrid as a two- to four-wire converter. The hybrid concept is shown in Figure 4-14. The block diagram of the generic dial telephone is shown in Figure 4-15.

With the development of touch-tone services, the push buttons selected combinations of two tones for each button. These tones were at first generated mechanically (in the form of chimes) and progressed to a series of elec-

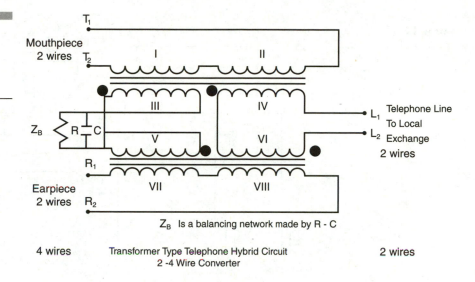

Figure 4-14

Transformer type telephone hybrid circuit 2-4 wire converter

Z_B Is a balancing network made by R - C

4 wires   Transformer Type Telephone Hybrid Circuit   2 wires
2 -4 Wire Converter

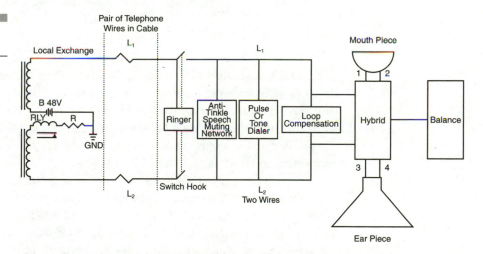

Figure 4-15

Basic telephone set

tronic oscillators. The electronic telephone is what we have today. A block diagram of such a telephone is shown in Figure 4-16. The action of the hybrid is synthesized by an *integrated circuit* (IC) chip.

The Local Exchange Office   The telephones in a local community all go to the LEX office by way of twisted-pair plastic insulated cables installed as buried plant or overhead utility poles. These wire pairs are connected to a crossbar switch that can be an electromechanical relay switch bank or a solid-state microprocessor-controlled switch bank.

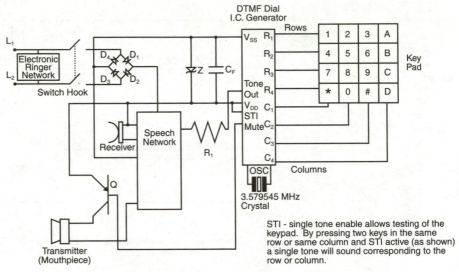

Figure 4-16
Electronic telephone
set

STI - single tone enable allows testing of the keypad. By pressing two keys in the same row or same column and STI active (as shown) a single tone will sound corresponding to the row or column.

Diode bridge consisting of D_1, D_2, D_3, and D_4 Protects against wrong polarity line connictions and feeds D.C. power to the speech network and DTMF generator. Z is a zener diode surge protector. C is a filter capacitor. Q acts as a muting control for the transmitter.

The LEX is connected to other LEX offices through an interoffice trunk line. Long-distance toll calls are connected to toll center offices that connect to inter-toll trunks. Formerly, the inter-office and inter-toll cable systems were copper, but most have been replaced by optical fibers. Regardless of the method, the telephone system network topology is the same. This interconnection topology is shown in Figure 4-17.

There are five principal classes of telephone system offices with the LEX office or end-user office. Long-distance telephone service can involve all five office classes interconnected by preferred and alternate routing. The hierarchy of a call between two parties using all classes of offices is shown in Figure 4-18. The classes of offices are supervisory in nature and have control over certain areas. The class 1 or regional office is the highest class and there are only 10 or 12 in the United States and two in Canada. There are more of the lower class offices with the class 4 office at the local level. Therefore, the class 5 office has the greatest number of offices.

The Telco Network   The telephone network in this country has contributed to the success of many companies using the services. The network has grown from many small regional exchange offices into a huge network providing wired communications for voice, data, and some video services. Actually, the system as it is today is a network within a network.

Figure 4-17
Telephone switching
topology

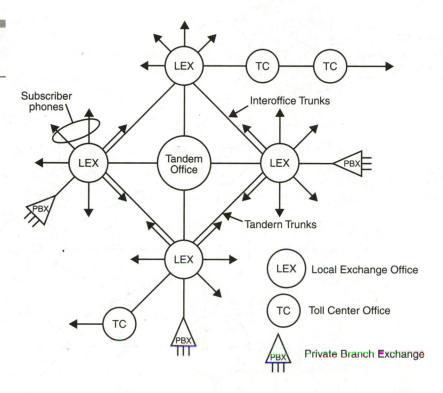

A supervisory network that the telephone companies use today to manage the service network is termed the SS7 network or Signalling System Seven. This system is a continuation of the *common channel inter-office signaling* (CCIS) network. The SS7 system uses a packet-switching protocol to send call-processing and status information that is essential to proper and accurate billing system charges to the subscribers and end users. Most of the switching is performed by electronic controlled switches referred to as 4ESS or 5ESS. The 5ESS switch is used when service demand is low and the 4ESS switches operate in a fully configured mesh-type network, providing many alternate paths needed to process a high volume of calls. Instructions setting up the call routes are sent by the SS7 signaling system to the 4ESS switches as the numbers are being dialed by the calling telephones.

Still, the largest providers of telephone service are the Bell system companies, often described as the *Regional Bell Operating Companies* (RBOCs). Other telephone providers operate in a similar fashion and all have access to each other's systems, thus providing proper seamless services to the businesses and general public.

The telephone network as we know it today grew from a simple copper wire-connected system to one consisting of coaxial cables, microwave radio,

Figure 4-18
Telephone system call
office hierarchy

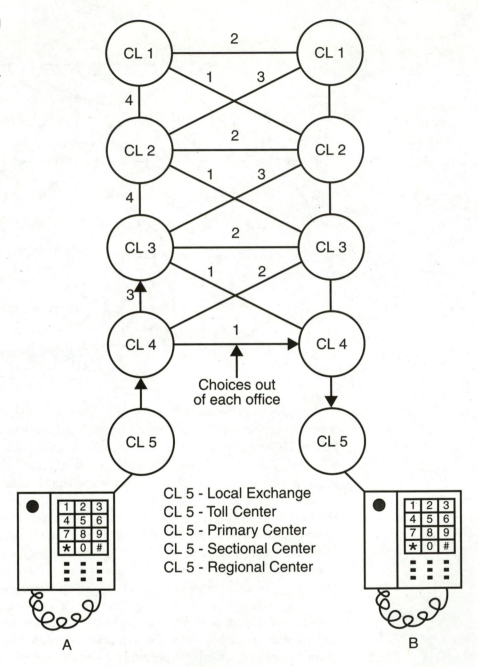

CL 5 - Local Exchange
CL 5 - Toll Center
CL 5 - Primary Center
CL 5 - Sectional Center
CL 5 - Regional Center

satellites, and optical fibers. These technologies are used for long-distance and trunk routing while the copper cables deliver telephone service the last mile.

Trunk Line/long Line Methods The trunk long-haul telephone services are processed over mostly optical fiber installed over the last 10 to 15 years. Switching functions performed by the 4ESS electronic switch are not simply connecting wires and cables together to connect the communication path. Since the telephone system converted to digital technology, telephone traffic consists of digital data words or bytes. Many calls are bundled together into digital transportation streams transmitted along trunk routes to the call destination where the bytes are stripped out, converted to analog speech, and sent to the called party's telephone.

It is interesting to note that early telephone long-haul trunking was done by coaxial cable. Analog voice signals were converted to higher frequency bands by single side-band radio techniques. One telephone one-way speech signal would occupy the upper side-band of a suppressed carrier and another the lower-side band. At the receiving end, the carrier was inserted and each side-band was demodulated.

Coaxial cables operating for long distances provided many telephone channels and cascaded amplifiers acting as signal repeaters were used along the cable run. Cascaded amplifier theory was put into practice and was the forerunner to cable television systems. The multiplexed signal scheme is shown in Figure 4-19. When no rights of way were available for cables, these signals could be converted to microwave radio frequencies and were transmitted the required distances. Converting back and forth between the cable signals and microwave radio techniques contributed to the transcontinental long-haul telephone system. Possibly, even today, some of these type facilities are still in use.

For quite some time now, the telephone companies have been installing large quantities of fiber-optic cable that is being used in long-haul trunk

Figure 4-19
Coaxial cable
12 voice

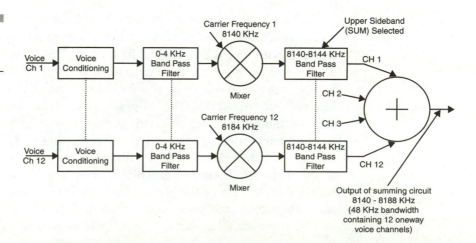

line applications. Also, the SS7 signaling network is nearly all optical fiber. Digital data communication traffic is transported mainly by the *synchronous optical network* (SONET) technology. The drop-add service at the various exchange offices is handled quite well by the SONET technology. Actual switching is performed by the *electronic switch system* (4ESS) by merely switching data bytes in an out-of-the-SONET data payload. A more detailed study of SONET technology appears later in this book.

Digital Telephone Methods

Shortly after the Second World War, the telephone industry converted to digital technology based on cost and system efficiency factors. This was a timely and appropriate decision to make. Since switching theory is actually a binary or digital subject, it is a normal extension of the technology to apply to the voice signal as well. Thus, the T-1 system was forthcoming. Because some telephone engineers use the word carrier to describe the message transportation method, the T-1 digital system is called the T-1 carrier, which is presently classified as *Digital System-1* (DS-1). Expansion of DS-1 speeds by time domain multiplexing (TDM) methods increased bit rates to DS-2 and DS-3 levels.

The T-1 Signal   The T-1 system digitizes 24 analog telephone channels into a data stream of bits. Since each telephone analog channel of four KHz is to be sampled, according to Nyquist's sampling theorem, each of the channels has to be sampled at 8,000 samples per second. Once each of the 24 channels has been sampled, a framing bit is needed to indicate the end of the frame.

Each sample of the analog data needs eight bits to accurately describe the signal amplitude. The net result is that 24 channels (one way) of the telephone service give a bit rate of 1.544 Mbps. This is shown in Figure 4-20. This method is described as pulse code modulation (PCM).

Also, since each telephone channel has a time slot in the bit stream, it is also referred to as TDM. The T-1 system can be carried on a four-wire system for full-duplex transmission. It might seem impossible that copper twisted-pair can carry a pulsed signal operating at 1.544 Mbps. This line rate is actually transmitted long distances as a carrier phase amplitude-modulated signal where several bits are contained by one phase and amplitude value. As long as the physical transmission wire or cable system can carry the carrier and preserve the modulation intact, the effective bit rate can be realized. Wider-band cables, such as coaxial cable and fiber-optic

Figure 4-20
Formation of T-1
systems (DS-1)

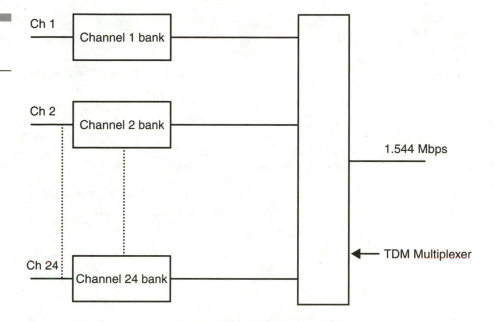

Each channel is sampled at 8000/second providing 8 bits/sample
which is 64000 bits/channel/second.
For 24 channels PLVS 1 bit (framing bit):

64000 bits/channel/sec * 24 channels/frame = 1536000 bits/sec
plus 8000 frame bits = 1544000 bits/sec = 1.544 Mbps.

cable, with higher bit rates can be transmitted. These higher bit rates or
T-1 carrier line rates are made by multiplexing the basic T-1 carrier.

Channel bank equipment either at the LEX or a *Private Branch
Exchange* (PBX) at various commercial locations can output higher order T
carriers. Present-day channel banks using *large-scale integrated circuit*
(LSI) technology work in full-duplex where the analog-transmitted speech
signal is digitized, compressed, and transmitted. The return digital signal is
also decoded, decompressed, and converted to analog, providing full-duplex
service. Channel banks that operate at higher speeds can stack basic T-1
carriers into higher order line speeds. For example, two T-1 frames consisting of a total of 48 channels will produce a line speed of 3.152 Mbps. This
method is illustrated in Figure 4-21. For short transmission distances of
under one mile, data can be sent over copper wires using a +3V and -3V
bipolar format called *Alternate Mark Inversion* (AMI). This format provides
improved signal synchronization due to the increase of transitions in the

Figure 4-21
Formation of T-1
systems (DS-1)

For Basic T-1 System: $\dfrac{8 \text{ Bits}}{\text{Channel}} \times \dfrac{24 \text{ Channels}}{\text{Frame}} = \dfrac{192 \text{ Bits}}{\text{Frame}} + \dfrac{1 \text{ Frame Bit}}{\text{Frame}} = \dfrac{193 \text{ Bits}}{\text{Frame}}$
DS-1

At a sampling rate of 8000 frames/second:

$$\dfrac{193 \text{ Bits}}{\text{Frame}} \times \dfrac{8000 \text{ Frames}}{\text{Second}} = 1.544 \text{ Mbps.}$$

For T-1 C: $\dfrac{8 \text{ Bits}}{\text{Channel}} \times \dfrac{48 \text{ Channels}}{\text{Frame}} = \dfrac{384 \text{ Bits}}{\text{Frame}} + \dfrac{10 \text{ Sync Bits}}{\text{Frame}} = \dfrac{394 \text{ Bits}}{\text{Frame}}$
DS-1C

At a sampling rate of 8000 frames/second:

$$\dfrac{394 \text{ Bits}}{\text{Frame}} \times \dfrac{8000 \text{ Frames}}{\text{Second}} = 3.152 \text{ Mbps.}$$

signal. The digital signal hierarchy developed from the basic T-1 system is shown in Table 4-3.

Expansion of Digital Telco   The telephone industry in both the United States and in Europe has expanded rapidly into digital technology. Basically, the T-1 digital carrier consisted of 24 analog telephone channels. Each channel is sampled at 8,000 samples per second into eight bits per sample, giving the basic DSO data rate of 64 Kbps. This DSO rate can support computer digital data. Therefore, a 24-channel T-1 system can be subdivided into, for instance, 18 voice channels (18 DSOs) plus five 56-Kbps (5DSO) channels, and three 9.6-Kbps data channels can be multiplexed to one DSO channel, therefore completing the 24 DSO channels. Using this method, commercial digital data services can be mixed into T-1 frames and transmitted to offices in various cities. Banks and financial institutions that require high-speed data communications use the telephone network facilities to carry their business data traffic intermingled with digital telephone traffic.

Packet-switched networks were developed in the United States in the 1960s. Such a network could carry both voice and data transmission. The original network consisted of nodes connected together over a wide area of leased lines. The nodes acted as switching centers. The terminal devices, known as *Data Terminal Equipment* (DTE) transmitted messages to other

Table 4-3

Formation of T-1
systems (DS-1)

Level	Number of Voice Channels	Bit Rate	Transmission Medium
DSO	1	64 Kbps	Copper twisted-pair cable
DS1	24	1.544 Mbps	Copper twisted-pair cable, coaxial or radio
DS1c	48	3.152 Mbps	Radio, coaxial cable optical fiber
DS2	96	6.132 Mbps	Radio, coaxial cable optical fiber
DS3	672	44.736 Mbps	Radio, Coaxial Cable optical fiber
DS4	4032	274.176 Mbps	Radio, coaxial cable optical fiber

DTEs. The early method operated as a datagram system sending small information packets called datagrams. As time went by, an agreement between the United States and Europe through efforts of the *International Standards Organization* (ISO) adopted the X.25 standards for packet-switched network interfaces. This concept used the first three layers of the *Open Systems Interconnect* (OSI). The OSI concept consists of a seven-layer protocol and is illustrated in Figure 4-22.

The main transport system for X.25 packet networks is done through leased telco lines with a direct dial-up backup. The packet networks developed for business data traffic work well in North America and Mexico. The DTEs used to be X.25-compatible and use the standard 15-pin connector. The X.25 packet also has to carry the appropriate flag, address, and control fields, followed by header and data fields. At completion of the three protocol layers of the OSI model, a frame-check sequence character is followed by another flag. Other network types requiring access to an X.25 packet network require the use of a piece of equipment called a bridge. The transmission speed depends on the specifications of the lines making up the physical layer and it is this type of communication service that can be developed by enterprising cable television systems.

Figure 4-22
Open system
interconnect 7
layer model

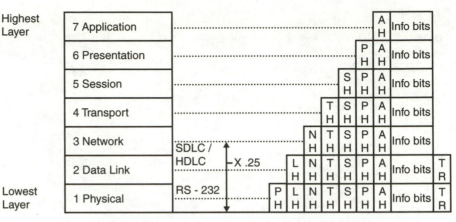

All Headers needed if all
7 layers are addressed.

Fiber-Optic Methods

As stated earlier, the telephone industry in the United States and Canada have installed thousands of miles of optical fiber, essentially replacing trunk lines and long-haul lines. Voice and data can be transmitted at extremely high speeds over long distances. Telephone traffic and data arrive at different nodes, all with various rates and without being synchronized. Thus, the *asynchronous transfer mode* (ATM) was developed to carry voice and data in the same packet. ATM accommodates the dispersion of voice packets and data packets better than a TDM or an ordinary packet-switched network. For TDM, each channel has a time slot assigned that makes up a frame. To intermingle voice and data into eight-bit octets, a voice channel will be assigned alternate timeslots with data. When no data is present for a timeslot, an idle octet is inserted. Packet-switched networks usually alternate between voice and data traffic. Both TDM and packet-switching can cause objectionable latency or delays for voice channels.

ATM-Sonet Systems ATM is more efficient since no idle octets are used and all timeslots can be used to carry data or voice traffic. This concept is shown in Figure 4-23. ATM makes more efficient use of bandwidth than other techniques and has the least delay times. The ATM cell contains 53 octets: 5 octets make up the header, and 48 octets compose the payload. An octet is an eight-bit group, often referred to as eight-bit bytes where two bytes would make 16 bits. Most system data conforms to multiples of octets

Figure 4-23
Comparison of TDM,
packet switching,
and ATM data
formats

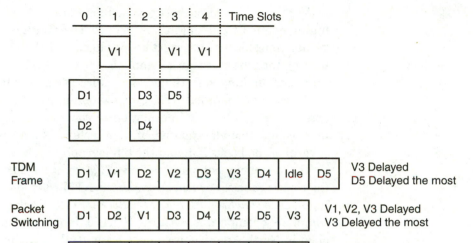

Conclusions: TDM favors voice traffic over packet switching.
TDM wastes time slots on idle.
ATM makes efficient use of bandwidth.

Figure 4-24
Format for an
ATM cell

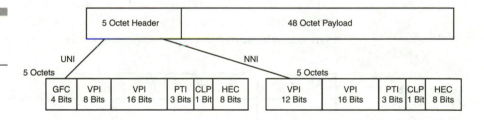

2 Types of Octet Headers UNI and NNI:

UNI Header	UNI Header
GFC Generic Flow Control	VPI Virtual Path Identifier
VPI Virtual Path Identifier	VC2 Virtual Channel Identifier
VC2 Virtual Channel Identifier	UNI User Network Interface
PTI Payload Type Identifier	PTI Payload Type Identifier
CLP Cell Loss Priority	CLP Cell Loss Priority
HEC Header Error Control	HEC Header Error Control
UNI User Network Interface	NNI Node Network Interface

that start with the eight-bit data word or byte and go on up to 64 bits. User data is parceled into the 48-octet payload portion of the cell while the header portion contains control, routing, and switching information. The ATM cell format is shown in Figure 4-24.

ATM is a form of packet-switched network technology where a cell is a highly defined fixed length unit of data. User data is not fitted into a particular timeslot. The user just keeps placing ATM cells on the transmission facility until the message is complete. The cell headers contain error control functions, making sure the header information is valid. The transmission facilities have installed ATM-type switching and routing equipment that acts on the cell header information to control the flow of the cells. For a more complete understanding of ATM technology, the reader should refer to several of the books listed in the bibliography.

The telephone companies carry ATM cells for customers using mostly fiber-optic cables. For longer distances, ATM cells are easily carried on the high-speed *synchronous optical network* (SONET) since the cell structure is completely compatible with the SONET requirements. SONET, as developed in North America, is compatible with the synchronous digital hierarchy SDH, which is a European standard. As shown in Figure 4-25, the common point of connection between SONET and SDH is the *Optical Carrier 3* (OC3) and SDH *Synchronous Transfer Mode* (STM) level, which is three times the OC-1 line rate. The SONET/SDH systems can handle the telco TDM *digital signal* (DS) formats with DS-3 the entry level. The DS-3 line rate at 44.736 Mbps (T-3 carrier system) can be fitted to the OC-1 slot with the necessary overhead.

There are many methods and specifications for transmitting data through communication networks. ATM is especially useful in gathering asynchronous data from various tributaries and forming data cells that can

Figure 4-25
SONET/SDH line
rate formation

Sonet	SDH	Line Rate (Mbps)	Multiplier
OC-1		51.84	
OC-3	STM-1	155.52	3 X OC-1
OC-9		466.56	9 X OC-1
OC-12	STM-4	622.08	12 X OC-1
OC-18		933.12	18 X OC-1
OC-24	STM-8	1244.16	24 X OC-1
OC-36	STM-12	1866.24	36 X OC-1
OC-48	STM-16	2488.32	48 X OC-1
OC-192		9953.28	192 X OC-1

be applied to SONET data payloads. In essence, non-synchronous data is transported by cells transmitted synchronously. Also, digital voice traffic latency or signal delay is not tolerated well. Customers do not want their voice conversations containing delays. ATM, as stated earlier, is the least objectionable as compared to a TDM or packet-switched method. Data, of course, can tolerate latency because receivers will wait until the whole message is received intact before transferring to the user for processing.

SONET Specifications The SONET system specifications and cell structure were developed by Bellcore and are considered as the high-speed toll road of the Information Highway. As developments occur in optical fiber transmission, data transfer rates have increased to the equivalent of 320 Gbps versus *Dense Wavelength Division Multiplexing* (DWDM) technology. The SONET transmission system has been shown to be like passengers on a commuter train. The first building block for SONET transmission is the STS-1 level, which consists of header information plus payload data. The STS-1 basic building block is shown in Figure 4-26.

The train similarity indicates the synchronous manner, or the connected together methodology of the SONET system. Synchronism must be maintained through the signal transport medium so the data slots can be

Figure 4-26

Synchronous transport signal (STS-1) basic building block format

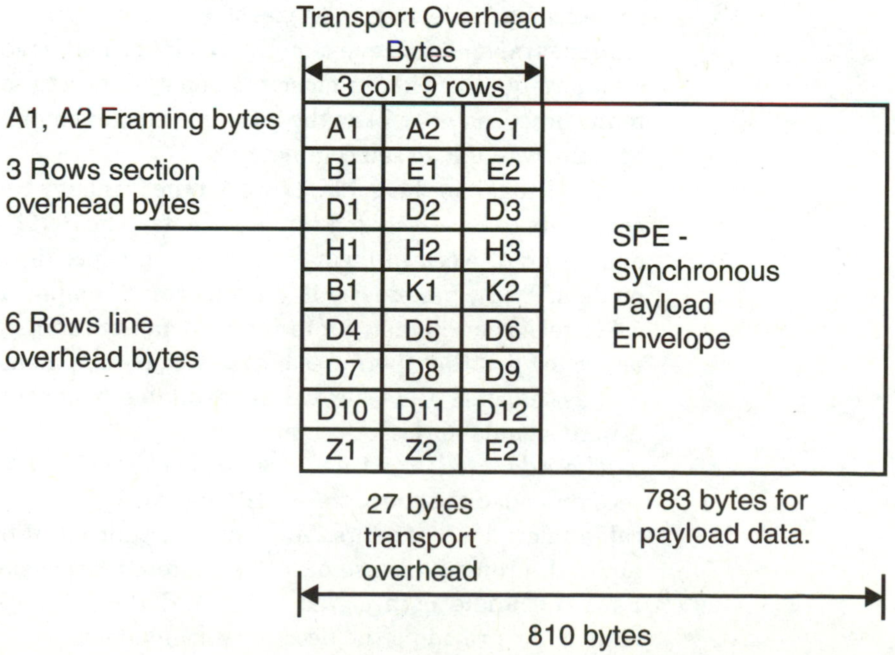

Transport Overhead Bytes

3 col - 9 rows

A1	A2	C1
B1	E1	E2
D1	D2	D3
H1	H2	H3
B1	K1	K2
D4	D5	D6
D7	D8	D9
D10	D11	D12
Z1	Z2	E2

A1, A2 Framing bytes

3 Rows section overhead bytes

6 Rows line overhead bytes

SPE - Synchronous Payload Envelope

27 bytes transport overhead

783 bytes for payload data.

810 bytes

unloaded and loaded with drop/add data riding the train. Scrambling the data bytes in the payload information provides enough signal transitions to aid in clock recovery. SONET-based systems can provide large amounts of high-speed digital transmission carrying voice, data, and video traffic. Readers interested in learning more about SONET should refer to Appendix C at the end of this book.

Digital Video & HDTV

The digital transmission of television in some parts of the country is operational and test data is being recorded and analyzed. Broadcasting in the digital domain has been studied and written about for a long time. The FCC finally accepted the transmission standard of 8 VSB along with all the baseband digital signal specifications necessary to implement service. Now broadcasting trials are underway. Several questions remain and answers should be forthcoming. Some of the questions are, how does a fading signal or multipath interference affect transmission and will many people need an outside antenna to obtain proper signal level?

From the consumer point of view, how much will a decent digital television receiver cost and will it have a flat, wall-mounted screen or be contained in a huge box the size of a piano? Also, will the broadcasting stations transmit in a one-channel HDTV format or several 4:3 aspect ratio standard-quality picture formats? Some observers in the field think that many broadcasters will do the latter since many more advertising slots will become available to sell commercials.

Also, what can the public expect in the form of a cable terminal with the capability to handle mixed signal formats? The SCTE and Cable Labs have done a lot of work in trying to resolve the specification of the best cable modem. Truly, this device is an important component needed to put the cable television industry into the next millennium. Many consumers still are concerned that their present television equipment will become obsolete and ask if some converter will be available to convert the digitally transmitted signals to NTSC format.

The *Advanced Television Standards Committee* (ATSC) of the FCC has recommended the use of the 8-VSB digital signal transmission scheme. The cable television operators, however, feel that QAM-64 is a better method and will promote its use on cable systems. Television sets somehow will have to handle both signal types and the IC chip manufacturers will respond by providing the necessary components.

NTSC & Digital Techniques

Digital television techniques have been around for nearly 20-odd years in the form of time-base correctors and video frame store devices. Most cable and broadcast television studios regarded such devices as simply black boxes because analog signals went in and came out. The digital technology simply operated inside the box. VCRs were also subsequently developed that used digital circuitry. The *Society of Motion Picture and Television Engineers* (SMPTE) then responded with standards for an all-digital television production studio. Thus, the frame store's switching and recording functions were performed in the digital domain.

Digital TV NTSC Digital technology was developed to operate with the standard NTSC video format presently in use today. The video signal is digitized by an *analog-to-digital converter* (ADC) by conventional means in which the resulting digital bit stream is processed and then converted back to an analog signal by a digital-to-analog converter. The processing might simply be a signal delay or a frame store function. Digital parameters, such as the sampling rate or the number of bits per sample that work with a 4.2-MHz bandwidth video signal, are handled within the device and are not under control of the user; hence, the phrase "black box" is used.

To convert the NTSC color television signal voltage to a digital data stream of pulses, an appropriate analog-to-digital converter is used. A basic ADC is shown in Figure 4-27a. In order to be able to convert back to an analog signal, the reverse process has to take place; namely, a *digital-to-analog converter* (DAC) has to be used. The block diagram for a DAC is shown in Figure 4-27b.

In order to convert analog signals to digital signals, certain rules dictate the sampling rate and the number of bits per sample. When transmitting

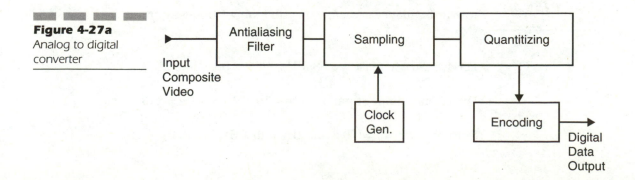

Figure 4-27a
Analog to digital converter

digital data through a communication channel, the necessary bandwidth and signal-to-noise ratio control the accuracy and quality of the digital signal. Resulting errors in the binary bits are referred to as the *bit error rate* (BER). This is an important measurement of digital signal quality. The input analog signal frequency is an important specification that controls the sampling rate and conversion to a digital data stream.

After much research and testing for the video signal, it was determined that the sampling rate should ideally be four times the chroma subcarrier frequency, or 14.3 MHz. For excellent signal resolution, a choice of 10 bits per sample results in a serial bit stream of 10 × 14.3, which equals 143 Mbps. The minimum bandwidth needed to pass this signal is equal to half this, or approximately 72 MHz. Clearly, this signal cannot be transmitted in the standard 6-MHz television band without some form of compression. Some of the rules and computations for NTSC video analog-to-digital conversion are shown in Example 4-1. Digital television signals are used in studios and post-production facilities and are not transmitted over the air. Therefore, the necessary bandwidth requirement is not an issue.

Figure 4-27b
Digital to analog converter

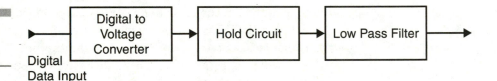

Digital
Data Input

Example 4-1
Establishing the bit rate

According to Nyquist's Sampling Theorom: The sampling

rate or frequency $fs = 2fb$ fb = upper frequency limit
                                        or bandwidth.

For $4 \times fsc$ where $fsc = 3.58$ MHz $4 \times 3.58 = 14.3$ MHz

For 10 bits/sample, total bit rate = 10×14.3 MHz = 143 Mbps

Since $fs = 2fb$, $fb = (fs)/2$ and $BW = fb$ $BW = 143/2$

$BW = 72$ MHz.

Digital HDTV   In this section, the process of forming a television image in the digital domain will be addressed. A system of progressive scanning is used, as opposed to interlaced scanning for analog NTSC television. A superior television picture will result by essentially increasing the number of pixels over the surface of the television screen. With NTSC 4:3, 525 line pictures, the horizontal resolution is a function of video bandwidth and is defined as the number of picture details that appear across three-quarters of the screen width. The number of horizontal details can be computed by multiplying the bandwidth in hertz by the horizontal scan time, giving 449 details. Three-quarters of this number gives 337 lines of horizontal resolution.

For vertical resolution, the total visible number of horizontal lines multiplied by 0.7, a utilization factor (Kell factor), gives 338, which is almost the same as the horizontal resolution. The pixel number corresponding to this vertical and horizontal resolution is simply the product of 449 times 338, for 15,761 total pixels in a picture frame. This calculation is a good milestone and is useful in comparing NTSC quality to other picture-making methods.

For color video transmission, the composite high-digitizing sampling rate at four times the chroma subcarrier frequency (four fsc) can be used. With NTSC, the digital signal can be broken down to the Y or luminance video signal sampled at four fsc and the I and Q chroma signal sampled at two fsc. When the Y signal and the chroma difference signals are digitized, Y is sampled at four fsc and the B-Y and R-Y at two fsc. This method is known as 4.2.2 sampling. The actual sampling is synchronized with the video signal chroma subcarrier, which is related to the horizontal scanning rate, thus establishing the sampling points on the video waveform. This subject is treated extensively in several of the references in the bibliography.

The important point is that the greater the number of picture-making elements of proper luminance and color, the better the picture quality. This allows for larger screen sizes in which any picture distortions are amplified. Each picture consists of a frame of pixels over the face of the screen with the position of each pixel fixed in location. The pixel locations on the screen form a matrix of vertical and horizontal pixels. The relationship between horizontal and vertical pixels, the aspect ratio, the frame rate, and the bit rate is shown in Table 4-4. It should be evident that digital compression has to be done so that digitally encoded television picture signals can fit into a realizable communication channel.

Video/Audio Compression   The compression method chosen for broadcast television is the MPEG-2 standard developed by SMPTE that compresses the high-bit rate digital television signal into a 19.5-Mbps compressed rate. This compressed rate is now transmitted as a 6-MHz band-

	H Pixel	V Pixel	Aspect Ratio	Frame/ Field Rate	No. of Bits	Bit Rate Mbps
HDTV	1920	1080	16:9	60	16	1990
NTSC-Quality	2640	2480	24:3	30	16	2147

width signal using 8-VSB modulation. The purpose of video or digital compression is to reduce the transmission bit rate to a value compatible to the transmission medium. Since NTSC video quality with a 4:3 aspect ratio requires approximately 150 Mbps to transmit pictures, a large value of bandwidth from the transmission medium will be required.

It is a well-known fact that successive pictures contain a large amount of redundant data. For example, an outdoor scene will have a nearly cloudless blue sky from picture to picture; thus, this same information is relatively constant from scene to scene. Many methods are used to reduce redundant information. The motion picture industry has been a big help in analyzing picture scenes and the way we see and perceive images.

Compression methods have been explored and studied in great detail. The MPEG-2 method produces excellent results and is the method recommended by the *Advisory Committee on Advanced Television Service* (ACATS). One of the earliest compression standards was known as MPEG-1 and was used as a digital storing method for video and audio programming at 1.8 Mbps. This method produces picture/audio quality similar to that produced by VHS cassette recording. MPEG-1 can compress 100 Mbps of sound or visuals (4095×4095 pixels) to a 1.8 Mbps bit rate, a 67:1 compression ratio.

As an extension of MPEG-1 came MPEG-2, and MPEG-1 is a subset of MPEG-2. Any MPEG-2 decoder can also handle MPEG-1 data. Thus, some compatibility between the two methods is maintained. As might be suspected, the MPEG-2 compression method is quite elegant and produces a *constant bit rate* (CBR). MPEG-2 can adjust to fast-moving picture components by using motion vectors that point in the direction of pixel movement.

Picture frames facilitate storing and/or switching functions. A bandwidth of six MHz, as specified by the FCC for television transmission, requires that the compressed 19.4-Mbps MPEG-2 signal will have to use some special modulation techniques. The 8-VSB method is the accepted method for broadcast television. Cable systems prefer and will most likely use QAM-64 for their modulation scheme. Whatever method is used, receivers for the public and cable subscribers will have to be able to detect both modulation methods. Some industry gurus are suggesting that most of the channel selection, detection, and decoding will take place within the cable modem, and the screen will handle and present the picture and audio signals.

Audio in the MPEG-2 format is also compressed. To cover the musical range of 40 Hz to 16 KHz, the sampling rate of 48 KHz is chosen, which is greater than the hyquist minimum of 40 KHz. At 16-bit resolution, a stereo (two channels left and right) will produce a bit rate of 1.54 Mbps. For a surround sound, or *Multi Television Sound* (MTS), a 4.5-Mbps rate is necessary.

Computer-type audio compression is used in producing CD-ROMs and is also used by the *digital broadcast satellites* (DBS) systems. These methods produce a properly compressed audio signal that can fit into MPEG-2 data frames. The audio and video data signals are formed into data packets that make up the program data stream. The MPEG-1 and 2 standards are presented in detail in some of the references in the bibliography. The digital data packets making up the MPEG-2 compression for both video and audio program content are shown in Figure 4-28.

The data stream for the MPEG-2 structure is simplistic in that it represents the basic format. There can be two audio packet formats consisting of mono audio or MTSC sound. Essentially, the various packets shown are sub-packetized with header information containing information about the data contained in the packet payload. The cable technicians, from a practical standpoint, do not need to be excessively concerned about the format, except to have a general knowledge of the digital data stream.

It is this data stream at 19.5 Mbps that will be modulated in 8-VSB by the TV broadcasting industry and transmitted on cable by QAM-64. The question remains, will the cable television industry receive off-air 8-VSB signals and carry them on the cable in the same format or demodulate them and retransmit in QAM-64? Word has it, that if consecutive channel 8-VSB

Figure 4-28
MPEG-2 program
bit stream

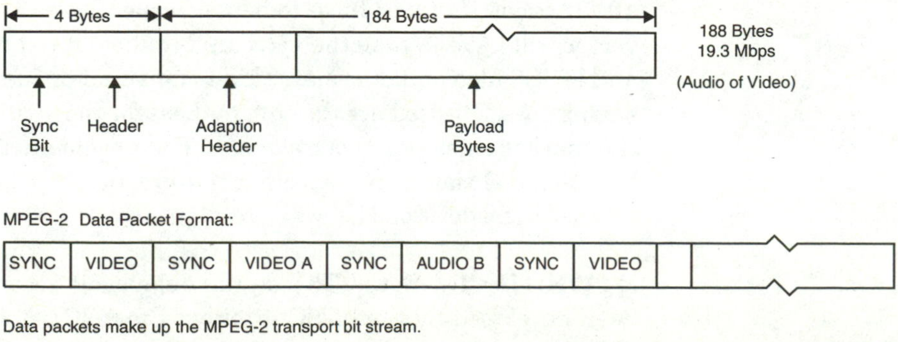

Data packets make up the MPEG-2 transport bit stream.

signals exhibit any interchannel interference, then it may be necessary to detect and retransmit in QAM-64. Ideally, cable operators should work closely with the broadcast industry to receive the MPEG-2 data stream of the broadcast stations signal via a fiber-optic connection that will preserve signal integrity. Today the cable operator can carry the station's signal in QAM-64.

Digital TV Transmission

The digital television set at this writing is not in existence as yet. Since the transmission system is still basically RF for television broadcasting, the digital television receiver will need the tuner to select the appropriate carrier frequency band. The tuner will produce an IF frequency containing the digital modulation. Detecting this frequency should produce the MPEG-2 bit stream, where the data will be decoded by an appropriate integrated circuit chip. Since it is expected that the digital television set will be able to process many digital television formats, several circuit chips will be needed to do so.

MPEG-2 Modulation Techniques   Digital television as broadcast in MPEG-2, 8-VSB, or QAM-64 requires a television set with digital processing circuitry to provide the picture and multichannel television sound. The cost, as projected by the consumer electronics industry, is regarded by many as excessive, which has infuriated much of the general public. Since digital television transmission renders the NTSC television set useless, it has been proposed that some type of converter should be made available to make the conversion. Undoubtedly, this converter will be extremely complex and hence expensive. Presently, no manufacturer has said they will have one available.

Thus, we have a market waiting for a product. It is clear that this converter will have to tune (select) the channel, demodulate the signal to recover the MPEG-2 and bit stream, and decode the bit stream. These are all functions that will have to be performed by a digital TV set. This converter will have to take the video and audio data and process it to develop an NTSC video/audio format. This is no small or inexpensive task. As the saying goes, "Blessed are the chip makers for they will develop a chip," since the market exists for this converter. The development of such a converter has been mentioned in the technical literature from time to time and many feel it will be developed.

DTV/HDTV TV Sets   The digital television receiver specification has been developed by the ATSC group of the FCC. As it is for most receiver transmitter links, the receiver follows the reverse process of the transmitter. Since the demodulated signal is a digital bit stream, the bit stream has to lock bit synchronization so the packet's video and audio can be found. The transmission layer has to be decoded as well as the transport layer.

The security control module or conditional access module contains the instructions for the receiver's de-encryption process. The transport layer provides the transmitted audio, video, and data packets, which are demultiplexed into their respective digital formats and sent to the video and audio processing circuits. Computer information consisting of games and computer displays are inputted through the IEEE P1394 I/0 port. The video format converter drives the video display with the information needed to form the images. A generalized block diagram is shown in Figure 4-29.

Repair and maintenance of digital television receivers will certainly be far more complicated than NTSC type sets. Highly trained service technicians using specialized test equipment will be required to service and maintain digital television receivers. It is hoped by many that HDTV receivers will be robust and require little if any service. This will most likely be true if the large-screen displays are developed that do not use high voltages and generate heat.

Broadcast HDTV The modulation method used to broadcast digital HDTV signals is the 8-VSB format. The block diagram describing this method is shown in Figure 4-30. From studying the figure, it is evident that this type of transmission is very different from what we are used to with the standard NTSC method. Digital signal modulation of an RF carrier takes place in many forms. Early studies of 8-VSB transmission have

Figure 4-29
Digital television
receiver

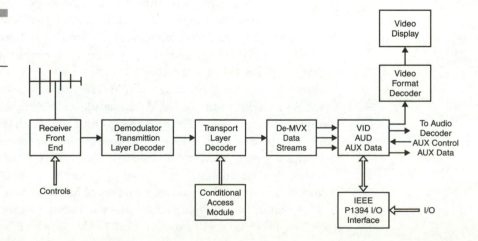

Figure 4-30
8 VSB transmitter
block diagram

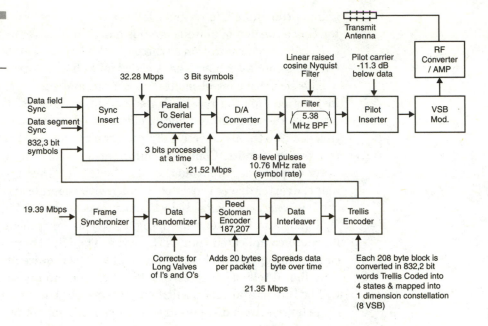

suggested that there is an adjacent channel interference problem possible, depending on the transmission method of the adjacent channel. At present, many cable operators are anxiously awaiting the field trials before making any decisions, but the cable industry feels that QAM-64 would be the modulation method of choice for digital transmission through the cable system.

QAM-64 Modulation QAM-64 modulation resulted from early digital transmission methods used in computer telephone modem development. This development resulted in higher modem equivalent bit rates such as 14.4 Kbps to 28.8 Kbps, which should be familiar to many of us. The QAM method of modulation uses both phase and amplitude modulation to achieve high bit-rate transmission.

As an example, a short study of QAM-16 will demonstrate the method of QAM. A block diagram of a QAM-16 modulator is shown in Figure 4-31. The incoming bit rate is divided by two into an I and Q channel. Two bits each are fed to the two- to four-level PAM converters in each of the I and Q channels where they are AM-modulated to an in-phase ("I") or to 90° shifted ("Q") carriers. The output signal is formed by summing the output of the I and Q channels. The constellation diagram for QAM-16 is shown in Figure 4-32.

This example was selected because it demonstrates the principles of operation for QAM digital modulation. QAM-64, which has been selected by

Figure 4-31

Block diagram of a
QAM-16 modualtor

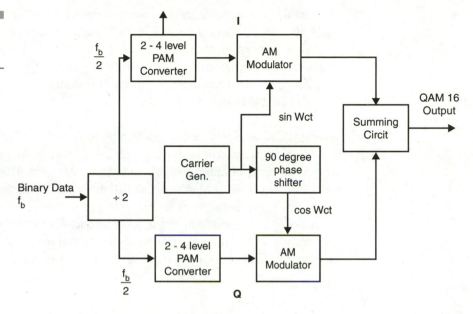

f_b = Bit Rate (frequency)

Figure 4-32

QAM-16 constellation
diagram

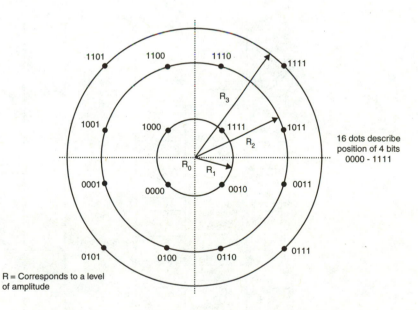

16 dots describe
position of 4 bits
0000 - 1111

R = Corresponds to a level
of amplitude

This diagram can be presented on an X - y oscilloscope by demodulating the I and Q signals
where the I channel is connected to the horizontal (X axis) and the Q channel is cinnected to
the vertical (Y axis). The oscilloscope's Z or intensity axis should be activated during each bit
position.

The oscilloscope constilation diagram should be exactly symmetrical with small clear dots
identifying the bits. Smearing dots indicate noise and nonsymmetry phrase errors.

the cable television industry as appropriate for transmission through amplifier cascades of a coaxial cable system, has more phases and amplitudes and operates similarly to QAM-16. QAM-64 has a more dense constellation diagram and is more difficult to analyze. Cable systems will require QAM-64 modulation equipment installed in its head-ends to transmit the digital television signals.

Digital Cable Head-end   The head-end for cable systems in the next millennium is going to have to be much larger and more complicated, requiring more maintenance, testing, and record keeping. The addition of digital services as well as analog services will exist side-by-side until the phase-out of analog services takes place.

Already many cable systems have added fiber-optic equipment in their head-ends that transports the combined signal to many nodes deep into the service area. Some of the larger and progressive cable systems have acti-

Figure 4-33
Head-end
configuration

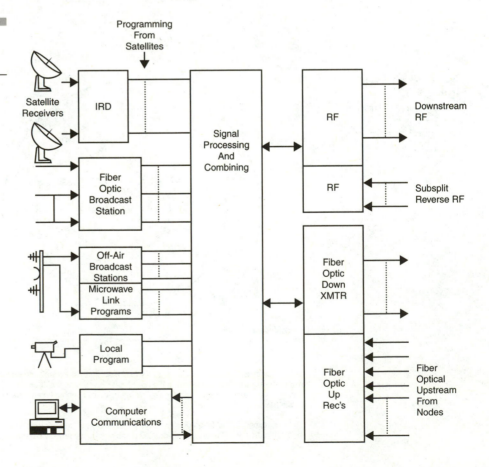

vated the reverse or upstream signal path back to the head-end. The network topology for the reverse system is often referred to as the reverse tree architecture since the forward system is known as tree and branch architecture. To solve the problem of reverse system noise combining at the head-end, upstream fiber-optic methods are used to divide the noise-source areas of the distribution nodes. Book references on this subject can be found in the bibliography.

The head-end as it exists today is the station where all signals exist. From there, upstream data signals are routed to their proper destinations. Signals from the various sources are routed to the various distribution nodes where they are distributed via short coaxial cable systems. A variety of test equipment and monitoring procedures are required to keep today's system head-end in proper operational condition. A block diagram of a type of head-end configuration is shown in Figure 4-33.

Of course, many types of head-end configurations can be found in systems today. Since reliability is now of utmost importance in cable systems, uninterruptable power supplies, signal monitoring, and alarms, as well as a stock of spare equipment, are all necessary. Above all, the level of training and expertise of a system's technical staff is of utmost importance.

CHAPTER 5

Subscriber Installation and Terminal Devices

The Subscriber Drop

The subscriber drop unfortunately has not had the attention and consideration it deserves from many cable operators. The nature of the cable television industry is partly responsible for poor and improper subscriber drop installations. At system turn-on or activation, the marketing and sales effort produces many subscribers ready to be installed. Most system operators do not have enough technical personnel to install the initial influx of subscribers, making it necessary to engage the services of outside-drop installation contractors. Traditionally, drop installation contractors go from system to system installing subscribers for various cable system operators. Unfortunately, supervision and quality control checks are often not done, allowing poor installations to occur. In many cases, the discovery of drop problems occurs long after the drop installation contractors have departed. Many systems find that a lot of correctional work has to be done to the drop installs to bring them up to proper specification. In many cases, this correctional work is unforseen and unbudgeted, causing unplanned expenses.

Early Installation Techniques

The aging of the drop install has been a problem for many cable operators. The improvement in the design of the "F"-type connector has undergone several major changes. The latest type, mentioned in Chapter 1, "Introduction," is of high quality and with proper installation should last a lot longer than the first generation connectors. Often, if subscribers have not complained of drop problems or no service calls have been performed, the original installation will be in operation even to this day. Systems expanding subscriber services by activating the upstream signal path quickly discovered serious noise and interference problems.

One of the main contributors is the subscriber drop. Many subscribers move the outlet from room to room, adding more cable and connectors to the internal system. The quality of components as well as the subscriber's work is usually substandard and contributes to return system noise. For many systems, the upstream frequency band is five to 40 MHz and is referred to as the subsplit return band. Noise generated in the subscriber drops are coupled into the upstream reverse amplifier cascade. Noise and interference is coupled at feeder-splitting points back into the bridging amplifier reverse module. Combining again takes place at trunk-splitting points and is transmitted through the return amplifiers to the head-end. An example is described in Figure 5-1.

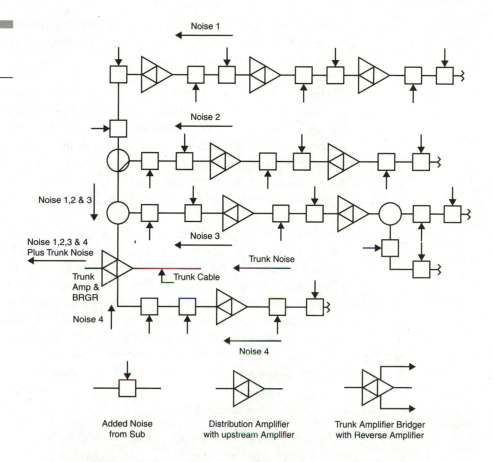

Figure 5-1
Up-stream noise combining

Since the early days, drop cable has undergone many improvements. Probably the most significant improvement has been in the increased shielding effectiveness. Present-day, high-quality drop cable has an aluminum foil wrapping the polyethylene foam dielectric and is covered with two woven aluminum braids woven over each other, followed by the polyethylene jacket. This type of cable gives a 90-percent shielding that amounts to over 100 dB of isolation. Essentially, this type of cable has three shields and is often called tri-shielded cable. Systems that experience significant noise build-up will have to budget a work project to prepare the return band for upstream signal transmission. A measurement of noise in the upstream band should be made at the head-end at points arriving from each trunk branch. Once a particularly noisy branch is identified, testing back along the trunk to the splits should be done to identify which, if any, is the most noisy point. Many systems have found that reworking the drops in the system using 90-percent shielded drop cable, new "F" connectors, and exterior drop hardware drastically improves the noise and ingress on the return path. Troubleshooting maintenance techniques in testing is covered in more detail in the next chapter.

Methods   Subscriber installations in many communities of single family homes or duplex apartments are aerial drops. The aerial drop, exposed to the elements of weather, deteriorate quicker than many underground drops. This is mainly due to wind, rain, ice, and snow stressing the drop cable at the points of connection, that is, both the utility pole and the house. An aerial subscriber drop is shown in Figure 5-2.

Also the effects of lightning can cause drop problems. If a lightning strike is close by, a power surge can take place or the cable grounding point can be affected to the point of burning or scorching the connector at the entrance point. If a signal outage occurs, then a service call will be requested by the subscriber and the cable technician will properly repair the problem. The effects of wind and the weight of ice can cause the cable to lose its concentricity, resulting in the increase in return loss and the shielding effectiveness of the cable. Systems experiencing such drop cable problems usually change to messenger cable, which gives added strength and support to the drop.

The material used in the subscriber drop was similar to that used by telephone companies, but it did not take the cable industry long to develop

Figure 5-2
Subscriber cable drop

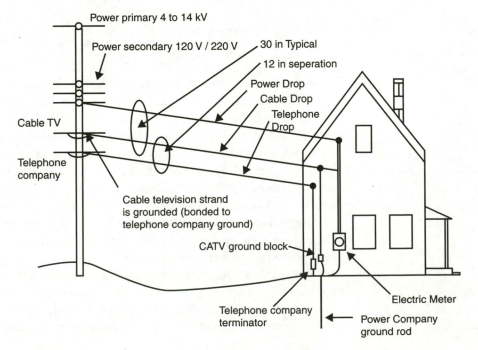

hardware and drop components that were more appropriate to its own requirements. The ease of installation with high-quality materials is the criteria that many progressive cable operators use regarding their subscriber drops. However, many cable operators have elected to use low-cost materials and poor drop installation contractors, which result in poor service to subscribers and a flood of service calls.

Since the points of connection at the utility pole and the subscriber's home are where the stresses appear, the hanging devices are extremely important. At the utility pole line, a span clamp is used to clamp to the steel messenger strand and provide a captive hook to fasten the drop wire. The drop cable containing its own steel messenger wire can be used by simply wrapping the stripped back steel strand tightly wrapped with several turns around the span clamp hook. For RG-6 cable that is non-messengered, a pre-formed galvanized wire forming a loop with the ends tightly twisted can be used as a cable grip. The twisted ends are coated with a sand-encrusted, semi-soft plastic covering. This device is referred to as a pre-form, which in its larger sizes is used in pole line construction, connecting strand together with splices or dead-end loops. The small version used in subscriber drops is the best method where non-messengered drop cable is used.

At the subscriber's house, a screw hook is used to fasten the drop cable. In many instances, installers fail to place this hook firmly into solid wood, causing the hook to pull loose and the drop to fall. This hook provides proper length to pass through the trim board into a corner timber. A twist in the hook eye prevents a cable using a pre-form to swing out of the hook. At this hook the cable is looped out and either stapled down the trim board or fastened with cable clamps down the side of the house to a grounded entrance block. This block contains a grounding connection for the ground wire connected to a grounding electrode. The grounding electrode is the electric utility ground rod or connecting wire to this rod. Cable operators should never loosen any clamps connecting wires to the ground point or rod. The proper technique is to add another ground clamp to the rod for the cable ground wire to attach. Drop connections for aerial drops are shown in Figure 5-3a and b for both the system point of contact and the subscriber's home. Proper hanging and placement of the drop wire is directly related to the life of the installation. The drop wire should not be drum tight nor hang too low. A good rule of thumb is to allow a sag at midpoint of one foot per 100 feet of drop cable at a temperature of 70° F. This is essentially eye-balled according to how the drop wire sags over the drop length. Most competent installers can often estimate the sag by how hard it is to pull the drop tight.

Some drop contractors and cable company installers often work as a single-person crew. Single-person installers often start in the house by placing the

Figure 5-3a
At house

Screw hook into
house corner

Messenger wrapped around
the hook and around the cable

Strip back to here

From pole

Drop wire

To subscriber

Figure 5-3b
At tap

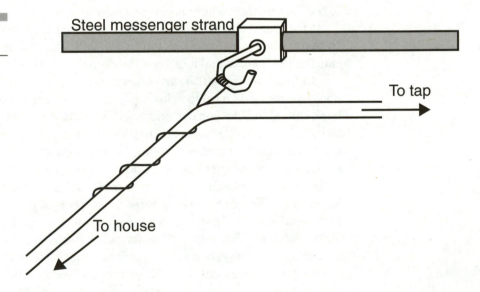

Steel messenger strand

To tap

To house

cable supply reel outside, drilling the access hole into a basement area or crawl space beneath the house, and threading the cable to the terminal location. At the terminal location, a connector is attached and the cable is fastened to the house timbers by either staples or cable slips. Staples should be either Monel or stainless steel composition. Cable clips are usually plastic fastened by aluminum or stainless steel nails. Once the initial inside work is finished, the installer can roll the excess cable back on the reel. The installer should now place the ground block near the hole where the cable comes through, install a connector on the cable, and connect to one terminal on the ground block, making a loop as shown in Figure 5-4.

Now the messengered drop cable is used for the actual hanging wire. The messenger wire should be separated from the cable a length approximately equal to the distance from the ground block to the house hook. Now a connector is installed and connected to the ground block through a similar loop and stapled or fastened to the side of the house with cable clips. The messenger wire is formed around the house hook and then back around the cable, previously shown in Figure 5-3a. At this time, the installer moves the cable supply reel to the pole, climbs the pole, and cuts a portion of the messenger loose, separating enough to wrap around the span clamp placed on the strand convenient to the tap port. Tensioning of the drop is done at this time by pulling the messenger wire through the hook and wrapping it around the hook and the drop wire, as previously shown in Figure 5-3b.

Figure 5-4
Aerial installation detail

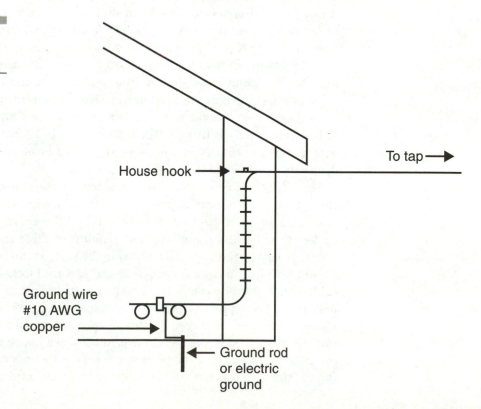

A connector placed on the drop wire and connected to the tap completes the pole work.

If not already completed previously, the installer should ground the ground block to the power ground electrode. If no ground is readily available, a cold water pipe exposed as a hose connection can be used, providing there is a direct connection to a metallic municipal water supply. In many instances, a subscriber can be connected via a plastic supply pipe or a driven well water supply. If it is necessary to drive a separate ground rod, a 5/8-inch copper clad rod is preferable to other types. It is often helpful and informative if the wiring inspector is consulted by the cable operator's technical department concerning any local codes and grounding-bonding practices.

Underground Methods   Many cable systems today are opting for underground or buried plants, following the trend of many telephone companies. Installing plant underground minimizes the problems of traffic hazards as well as weather hazards. In many urban residential subdivisions, it is required that all utility services be placed underground.

Underground drop wires have been in existence for a long time. Early underground drops were placed in a slit in the ground made by a flat spade or lawn edger. As the need appeared for faster and deeper drop installations, a variety of powered machines appeared. Present-day machines plow drop cables to depths of a foot at rapid rates approximately 15 feet per minute through average soil conditions. Soft, wide tires allow these small cable plowing machines to cross lawns with a minimum of restoration problems. At the pole, the cable is connected to a tap port and is either stapled or fastened to the utility pole. The bottom eight to 10 feet should be covered with a molding to protect the cable at ground level. A correctly installed drop cable at the pole is shown in Figure 5-5.

Usually the cable is plowed from the base of the pole to the house where the ground block is located. If a sidewalk is located between the pole and the house, a section of steel or "PVC schedule 40" conduit should be pushed under the sidewalk. An appropriate amount of cable should be left at the pole to reach the tap port and dressed to the pole. At the house, enough cable should be left to form a drip loop at the ground block. Direct burial drop cable with a flooding compound should be used for such underground drop installations. For present-day cable systems offering service in the 800-MHz range, only RG-6 or larger 75-ohm drop cable should be used. A typical underground drop connection at the subscriber's house is shown in Figure 5-6. Many new home developments are served by underground cable plant in pedestals or vaults, requiring drops to be placed underground as well.

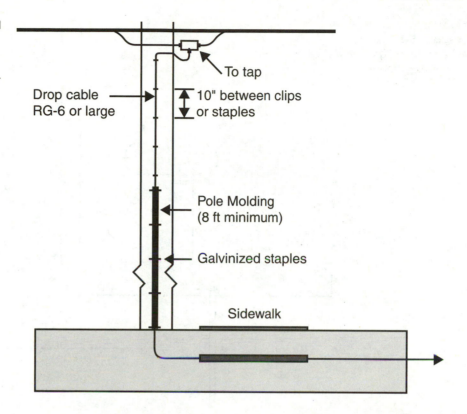

Figure 5-5
Underground drop
shown at pole

To tap

Drop cable
RG-6 or large

10" between clips
or staples

Pole Molding
(8 ft minimum)

Galvinized staples

Sidewalk

Splitting & Coupling Most cable subscribers require more than one cable outlet and many subscribers have a whole interior cable distribution system. Often new homes are pre-wired using either a loop-through or home-run system. Both types are shown in Figure 5-7. Multiple outlets require either signal splitters or directional couplers to distribute the input signal level among the outlets. Some cable operators feel that the loop-through system saves cable and hence cost, while other operators think that more directional couplers offset any savings in cable costs. The choices are not often clear-cut and often depend on the expertise of the installer technician. However, ideally all subscriber outlets should have nearly the same signal level and enough levels to operate the TV set or converter terminal. Some cable operators offer pre-wire services to developers or subscribers with costs rebated at activation. Cable operators doing it this way have some control over the quality of the installation. Unfortunately, many subscribers elect to extend or rework their interior cable distribution system using inferior components often improperly installed. This results in

Figure 5-6
Underground drop to
subscribers house

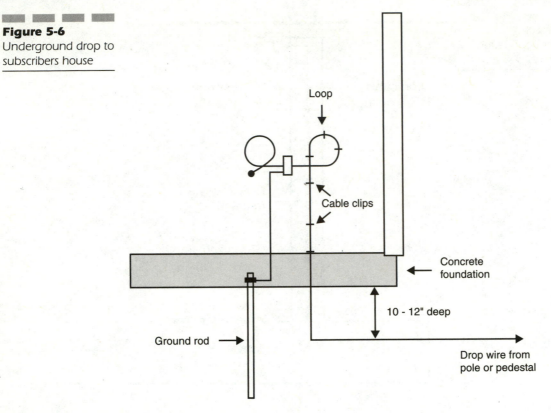

Figure 5-7a
Four-outlet loop-
through using
directional couplers

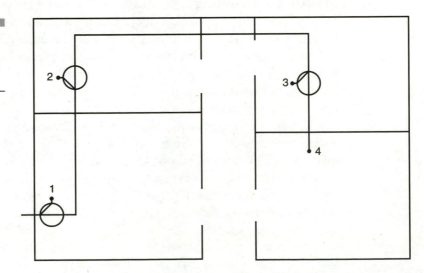

4 Outlets Loop Through using Directional Couplers

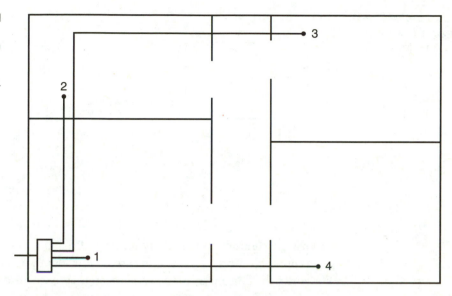

Figure 5-7b
Four outlets homerun
type using one four-
way splitter

poor service, for which the cable company is blamed. Systems that have a large number of poor subscriber installations can experience noise and ingress problems in the upstream return path.

Most subscribers elect for usually two or three cable outlets covering a family room, bedroom, and possibly a kitchen-dinette area. Signal-splitting can often be conveniently made outdoors where the drop connects. Many vendors of drop equipment offer two, three, and four outlet outdoor signal-splitters with a grounding connection. This type of connection does not need a separate ground block since these devices are fitted with a ground wire connection. A splitter of this type is shown in Figure 5-8.

Grounding and Bonding   Two of the most important aspects of the subscriber installation are grounding and bonding. Proper grounding and bonding protects the drop as well as the subscriber's equipment from the effects of lightning. In lightning-prone areas, often soil conditions do not provide adequate ground conduction necessary for a good earth ground. Often several ground rods screwed end to end are necessary to obtain a decent earth ground. Some suburban residential areas have a metallic water supply system, which can act as a good ground for the telephone electrical and cable system. Electrical utilities are grounded using ground rods or to the water system, provided it is metallic. However, the proper ground electrode for cable systems is the electrode used by the utility companies, which is either a rod or water system.

Figure 5-8
Outside grounded splitter

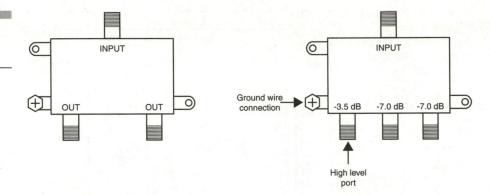

Ground conduction is usually measured using a device called a "megger." Two copper ground rods are separated a few feet apart and are driven to the same depth. The instrument is connected to each ground rod, which then places a difference of potential between the ground rods measuring the resulting current. The ground resistance is read from the instrument scale or can be calculated by dividing the test voltage by the measured current. Most instruments of this type provide a direct reading of the ground resistance between the two rods. The proper measurement of ground conduction is the ratio of the ground resistance to the distance between the rods given in ohms per foot. Often the desired result is 25 ohms per foot. If need be, a longer rod or a rod with a larger diameter may be necessary to achieve this result.

The conductor size used to ground the cable service to the grounding electrode is #14 AWG copper wire. This size is the minimum that should be used and many cable operators use a larger size of #12 AWG solid copper conductor with PVC insulation. The insulation offers some added strength to the wire and protection against corrosion. This type is usually readily available from either cable drop supply vendors or electrical supply companies. Such suppliers usually carry ground clamps for a cold water pipe connection or ground rod clamps and split bolt connectors that are used to connect to another ground wire. Several type of these devices are shown in Figure 5-9.

The usual installation procedure for proper grounding is to determine how the power neutral is grounded. If the power ground electrode is a driven ground rod, a ground rod clamp can be placed on the rod and a #14 or large copper wire can connect the cable ground block to the clamp. If the top of the rod is not available, the cable ground wire can be attached to the power ground wire using a split bolt connector or the wire can be connected to the power metallic entrance box using a spade lug and screw.

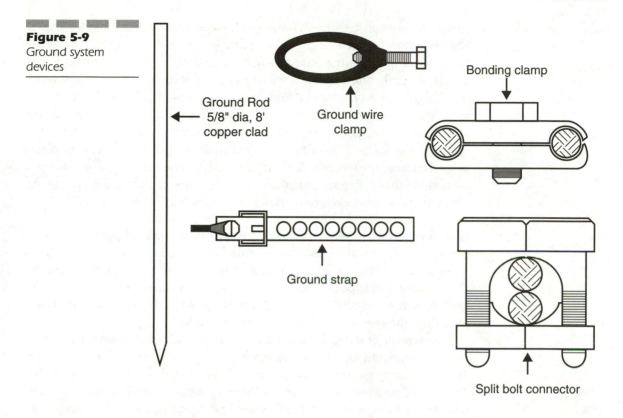

Figure 5-9
Ground system devices

Ground Rod 5/8" dia, 8' copper clad

Ground wire clamp

Bonding clamp

Ground strap

Split bolt connector

At the poles, any bonding is done using bare #6 AWG copper wire using strand-bonding clamps connected to the cable system strand and the telephone company strand. The bonding procedures and specifications for an aerial plant are usually spelled out in detail in the pole attachment agreement between the cable operator, the telephone company, and the power company. Proper grounding and bonding is extremely important to the cable operator as well as the utilities. The cable operator's technical staff should instruct all cable installers on proper procedures for grounding and bonding the distribution plant and the subscriber's home.

Several large *multiple system operators* (MSO), as they are known in the cable industry, are presently or will be offering telephone service in the near future. It is common knowledge that telephone systems carry their own power. Thus, a loss of electrical power does not cause the telephone system to fail. Cable systems offering telephone service are going to have to provide all signals and power to the telephone terminal in the subscriber's home. This means the drop cable is going to carry electrical power through the install cable path to the telephone set. For proper protection of power surges, a surge protector will have to be placed at the ground block location. This necessitates a cable communication entrance box housing to contain the ground block, surge suppressor, and possibly a signal splitter. Fortunately, several vendors offer a variety of such products to the cable industry. Also, for subscribers not taking any services requiring the upstream path, the placement of a filter in this cable entrance box will prevent any ingress and noise getting into the return system. The use of an entrance box makes a nice-looking install and keeps the drop equipment and connectors protected from the weather. This adds to the system reliability necessary to compete with other communication providers. An example of an outside residential entrance enclosure is shown in Figure 5-10.

Quality Control   Most cable system operators perform little or no quality control checks on subscriber installations, regardless of whether they were done in-house or by a drop contractor. The reason, most likely, is due to extra cost and expense. This, of course, is a serious mistake that can come back to haunt the cable operator. After all, it is the drop connection that is providing the service the subscriber is paying for.

Two schools of thought exist for checking out subscriber drops. One is to have a technician follow around the installation crews and check on every drop. This is, by far, the most comprehensive as well as the most expensive. The second method is to spot-check the drop installation. Finding a problem with the quality of the work will point the finger to the crew responsible, which is the type of check preferred by most cable operators.

The first thing a cable operator must do to obtain quality drop installations is to make sure that the installers, either in-house or contractors, have proper training in the way the cable operator wants the work to be per-

Figure 5-10
Cable
communication
entrance enclosure

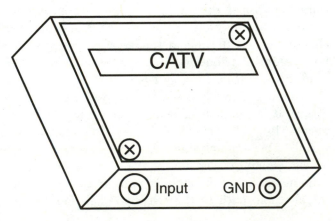

formed. This includes specifying the drop material and the methods for doing the job indicated by the cable operator. Cable operators should start quality control checks on the installations either on a complete or a spot-check basis soon after the work commences. This will nip any problem areas in the bud.

The cable operator should keep accurate records recording which drops were inspected as well as noting any problems and comments. Forms for these inspections can be developed and printed in tablets. The inspector can then check the items listed as being per drop material and installation, as shown on the form. At the end of the day, the inspection form information can either be filed or entered into the technical file computer database. By using a hand-held computer in the field at each inspection location, data pertaining to the inspection can be entered directly. This information can be accumulated on diskette and either filed or compiled in the technical database. The cable industry should, if it not already has, obtain appropriate software to aid in proper record keeping. Essentially, the information entered on a form or on a computer should be the same. An example of an inspection form is shown in Figure 5-11. Whichever method is used, the quality control information at initial turn-on is important for maintaining the drops and subscribers' service, as well as studying the effects of the drops as the system ages.

Drop Aging & Deterioration

The aging of the subscriber's drop is definitely related to costs and expenses. It is a known fact in the cable television industry that high-quality drop installations may cost slightly more in the beginning but will definitely last longer and be relatively problem-free. Unfortunately, most plants and equipment under the conditions of weather and seasonal temperature changes deteriorate with age. Thus, the outside subscriber's drop has essentially a limited life span that varies with the quality of the initial installation. For many systems, the life of a drop installation varies from eight to 12 years, which is also often the length of the system financial loan. Therefore, the buyer of a 10-year old system should evaluate the condition of the subscriber drops before making any financial offer. A new prospective owner of a 10-year old system should plan on replacing the drops as part of an overall system upgrade.

Corrosion & Moisture   Metallic corrosion due to moisture is one of the concerns of drop deterioration. Iron rusts, aluminum oxidizes, and copper turns green; all are forms of corrosion. Metalurgists have studied the

Figure 5-11
Installation quality
control checklist

INSTALLATION QUALITY CONTROL CHECKLIST

OUTSIDE

ADDRESS OF SUBSCRIBER:
STREET NAME / NO. _____

AERIAL / U.G. _____ LENGTH OF DROP (EST) _____ FT

MIDSPAN / DIRECT _____ DIRECT POLE TO HOUSE (Y/N) _____

GROUND ELECTRODE:
SEPARATE ROD   Y _____   N _____

ELECTRIC UTILITY:
GROUND ROD _____ CONDUIT _____
METER BOX _____ WIRE _____

WATER PIPE CONNECTION _____

GROUND CONNECTIONS:
TIGHTNESS _____ WIRE PLACEMENT_____

GROUND BLOCK "F" CONNECTORS:
APPROVED CONNECTORS (Y/N) _____WEATHER BOOTS (Y/N) _____
USE OF SEALING COMPOUND (Y/N) _____

SIGNAL LEVEL AT GROUND BLOCK:
LOW CH # _____   HIGH CH # _____

INSIDE

NUMBER OF SPLITS _____ SPLITTERS / COUPLER_____

SIGNAL LEVEL AT FARTHEST POINT FROM GROUND BLOCK:
LOW CH # _____   HIGH CH # _____

CONVERTER / SUBSCRIBER MODEM:
TYPE USED_____

SUBSCRIBER'S COMMENTS ON SERVICE AND WORK DONE:
_____
_____
_____
_____

NAME OF TECHNICIAN   _____   DATE _____   TIME_____

effects of metal corrosion and the effects of dissimilar metals in contact. Throughout the years, the metal industry has provided corrosion-resistant metal alloys for manufacturers to make metal components for the communications industry. Aluminum alloy amplifier, tap, and passive housings are all used in the outside cable plant. Ground blocks are made of aluminum alloy and are fitted with plated brass cable connections. The latest version of the "F" connector is made of high-quality metals and contains an interior O-ring to keep moisture from entering the connector from the cable end. The manufacturing industry has responded with high-quality drop installation components and cable systems should use them. Splitters and couplers are made from corrosion-resistant aluminum alloys and some use stainless steel. All are sealed against outside moisture. Still, if some of these enclosures are not hermetically sealed, moisture from the outside air can carry moisture to the inside and water can condense, causing interior corrosion. If such components can survive 10 winters of rain, ice, and snow, this amounts to a 10-year life span for drop components. As subscribers move from one location to another causing subscriber churn, a cable operator should use this disconnect/reconnect cycle to upgrade the drop installations.

Drop Cable Stress Damage   The continual stretching and relaxing of drop cable in the wind causes it to deteriorate. Also, temperature effects of expansion and contraction add to the wear on the drop cable. Since this cable is a transmission line with a 75-ohm characteristic impedance, it must retain its characteristics or the signal it carries can deteriorate. The center conductor must remain in the exact center and an equal distance from the shield, thus maintaining its coaxial condition necessary to maintain its characteristic impedance. If the source (driving) and the load impedance are matched at 75 ohms, the cable has to be 75 ohms over the frequency range in order to maintain the matched condition. Most of us know when the load or source impedance is different than the cable characteristic impedance, a mismatch occurs and signal reflections take place. These reflections cause the subscribers' pictures to have ghosts, which is an objectionable picture impairment.

Hanging cable is usually kept at a prescribed sag, which allows for expansion and contraction. The cable grips mentioned previously called preforms keep the hanging stresses from affecting the connector at the tap, which can also be accomplished by using messengered drop cable. Still, the nearly constant motion of aerial cable can cause stress cracks in the outer jacket, allowing water and moisture to enter and start corrosion of the cable shielding. Proper attention of the drop at a service call can head-off problems later on. A few more minutes spent by a competent service technician observing the condition of the drop installation should be a company rule.

Lightning & Electrical Effects   The effects of lightning on subscriber drops can be catastrophic. Systems that usually have lightning problems are in what is known as lightning-prone areas. Various sections of the United States are more prone to lightning than others and a map showing different areas' rates of thunderstorms is shown in Figure 5-12. Such a map is called an isokeraunic map. Obviously, the southern and southeastern states are more likely to experience lightning problems. Often these same locations have sandy soil and maintaining a good earth ground is a problem concerning not only the cable operator, but also the electrical utility company and the telephone company as well. As previously discussed, a good ground at the ground block is necessary to protect the subscriber's property as well as the cable service.

Cable drops that have been affected by a direct or nearby strike can exhibit scorching around the ground block. If this happens, the ground wire and connections should be checked and possibly the whole ground block and the associated cable connectors should be replaced. All the subscriber outlets should be tested for signal levels in case some of the components found in the splitters or directional couplers have been burned out. Sometimes the effects of lightning show up at a later date, resulting in a call back. Since the subscriber's television sets and converters are connected to the power system and power system ground, proper ground bonding should be done to protect any currents due to lightning to flow in the home wiring. How this

Figure 5-12
Annual thunderstorm days

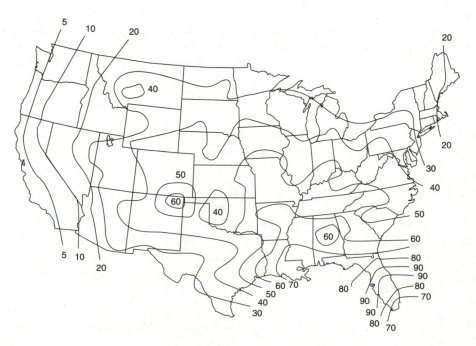

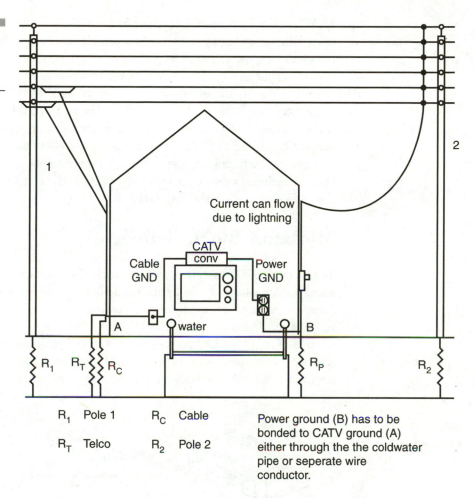

Figure 5-13
Ground bonding for protection from lightning

Current can flow due to lightning

CATV conv

Cable GND

Power GND

A   water   B

R_1 R_T R_C R_P R_2

R_1	Pole 1	R_C	Cable
R_T	Telco	R_2	Pole 2

Power ground (B) has to be bonded to CATV ground (A) either through the the coldwater pipe or seperate wire conductor.

happens can be shown by considering Figure 5-13. The 75-ohm to 300-ohm matching transformer required by older television sets is easily affected by lightning and often has to be changed after particularly heavy lightning. In many cases, these transformers protect the set itself by "taking the hit."

Subscriber Converters

Subscriber converters have been through many generic versions throughout the years. The first use for a subscriber converter was to enable a television set to tune to a channel it could not receive otherwise. Early television sets were basically a 12-channel set offering channels 2 through

13. When the FCC opened up the UHF television band, television sets made after that time became fitted with a UHF channel selector, while earlier television set owners had to purchase a separate UHF converter. As mentioned earlier in Chapter 1, cable operators provided 12 channels on a cable system with an upper frequency limit of 220 MHz. UHF stations were translated down to the band of 108 MHz to 170 MHz, which could carry nine mid-band television carriers. The translation was done at the cable system head-end by adding these nine channels to the already established 12 channels, which gave a system 21 total channels. A subscriber converter then translated these channels back up to UHF channels to be selected on the television set's UHF converter.

Midband Block Converters

This type of UHF converter was considered a mid-band block converter that translated a block of mid-band frequencies up to UHF channels. It used a fixed frequency local oscillator that transformed the mid-band channels up to UHF standard channels. Such a converter is shown in Figure 5-14. Converters of this type had no controls or On/Off switches and remained active as long as the plug was in the wall. When pay or premium programming became available, the mid-band channels were used to carry this service.

Figure 5-14
UHF block converter

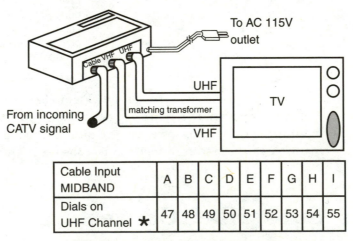

Cable Input MIDBAND	A	B	C	D	E	F	G	H	I
Dials on UHF Channel ★	47	48	49	50	51	52	53	54	55

★ Other channels may be translated to other UHF channels.

Usually one or two premium channels were grouped together and separated from the UHF channels by a blank channel. This converter also enabled the use of a remote control, yet when this type of converter was in use, few television sets had remote controls.

Signal Security Methods   When premium or pay channels were added to cable systems, some means of preventing the unauthorized use of them was needed. Since a cable system grouped pay channels consecutively, a signal-delete filter (trap) was devised to remove the premium channels from the subscriber's drop. This trap appears in a barrel-type configuration with a male connector on one end that was screwed into the tap port, and the drop was then connected to the other end. Since this type of trap was used to deny the premium channels from a subscriber, all non-takers of the premium channels had use of this trap.

For systems without many premium programming subscribers, a lot of traps were used. This, of course, was an expensive choice. A so-called "fix" to this situation was to insert an interfering carrier into the premium channels that would cause intentionally severe picture distortions. Placing a sharp notch filter in the drop, usually at the subscriber's tap port, would delete the interfering carrier, allowing clear reception of the premium channels. This type of trap was known as a positive trap. It was needed only by premium subscribers and thus a system with few premium subscribers did not need to use many units and would pay few expenses. If 50 percent of a system's customers were premium subscribers, it would made no difference in costs, regardless of which method was chosen, since both the positive and signal-delete types of traps cost nearly the same amount of money. Figure 5-15 shows what these traps look like.

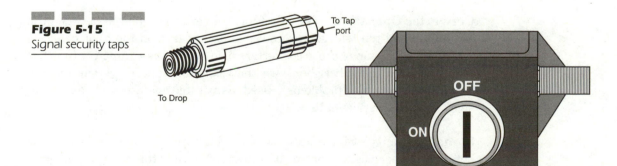

Figure 5-15
Signal security taps

A system audit of the number of active premium subscribers a system had could easily be performed by viewing the system taps on a drive-by basis. Eventually, people stealing the premium service were found by systems who discovered a look-alike dummy trap that did not function. Cable television supply vendors quickly developed a trap security shield that enclosed the trap and required a special tool to either install or remove it.

Signal Quality Factors The use of converters and traps did have some affect on picture quality. The block converters could drift a bit and thus the subscriber would have to readjust the fine tuning on the television set. Leaving the converter powered constantly helped the drift problem. Television sets at the time had poor noise figures for the UHF tuner and since the block converter translated the mid-band channels to the UHF tuner, slight snow could be seen by subscribers viewing these channels. Negative-type traps had to be made with steep edges above and below the pass band so as not to remove any picture information from adjacent channels. Positive traps, while removing the interfering carrier, would often remove some sideband signal components, causing some amount of picture impairment.

The Selectable Converter

An improvement to this process was the selectable converter that allowed all program selections to be made by the converter channel selector. The channels selected on the converter were converted to either channel 3 or 4 on the VHF channel selector. This selectable converter began many disagreements between the consumer electronic industry and cable operators. Unfortunately, many of these converters were poorly made and caused cable systems a lot of trouble.

Converter Operation The selectable converter was placed between the cable drop and the television receiver and appeared on top of or near the television set. Some of these converters also had a fine-tuning control that often confused the subscriber with the television tuning system and the converters. Mistuning problems caused many cable operators a lot of expensive subscriber service calls.

Signal Security Signal security with this type of system was done as before by either positive or negative traps. Actually, this type of converter did little for the cable operator and in many instances caused more problems than it solved.

The Programmable Converter

The electronic programmable converter was introduced when signal scrambling methods were developed. Signal scrambling of the premium pay channels was done at the cable system head-end and it was the job of the converter to unscramble the signals at the subscriber's television set. The first generation of this type of converter used a "PROM" chip with the decode logic perminately burned into its circuits. Cable operators could order from suppliers the pre-programmed chips or could program their own using a PROM programmer often referred to as a PROM burner. It did not take long for private chip operators to appear, allowing unauthorized use of a cable system's premium pay channels. This converter contained the descrambling circuits that would descramble channels authorized by the PROM chip. Again, channel selection was performed by the converter and translated to a low VHF channel, usually channel 3 or 4. This made the television set's remote control useless. Since many subscribers wanted remote control, cable converters with remote controls became available. Now subscribers had two remote controls and one was useless. The relationship between the consumer electronics industry and the cable system operator as well as the subscriber had not improved.

Signal Scrambling   Signal-scrambling methodology was in the beginning quite simple and effective. Removing the horizontal synchronization pulse caused television sets to lose synchronization, destroying the picture, yet the audio signal was not disturbed. In order to remove the horizontal synch pulse, the scrambling circuit had to find it. This is actually quite easy circuit-wise and at each sync pulse interval a circuit shorted it out to ground level (zero volts), thus removing its presence from the video signal. Actually, this was done at the video IF inside scrambler at the head-end. In the converter, the reverse procedure was performed on command from the programmed PROM, thus restoring the synch pulse. One of the problems with this converter is that they could "walk," meaning it could be moved to another location and be used with any television set.

Prom Programming   The authorization and security control remained with the PROM. By using the PROM programmer, some cable systems tried to use more complicated authorization codes. Still for many systems, the signal thieves just became more clever. Many manufacturers of cable converters worked with cable operators to try and keep up with the theft of service problems.

The Addressable Converter

The addressable-type converter seemed to solve some of the security problems. With the addressable converter, an exclusive code for the subscriber was programmed into its PROM. Each converter had to have this code refreshed at predetermined intervals or the converter would shut down and cease operating. Also, the information for the selected channels to decode was transmitted on the downstream data signal along with the address data. This data stream was transmitted just above the FM band using FSK-type modulation. This type of converter allowed a subscriber to call the cable customer service office and order a change in pay service. The controlling computer, which was also usually a part of the billing system, would select the subscriber code and change the decode instructions for the desired channel and delete any others from the service. This, of course, was a savings to the cable operator because a technician did not have to physically go to the subscriber's home since the changes would be made in the office.

Converter Initialization   Converters shipped from the factory to a cable operator had to be initialized before being placed in service at a subscriber's home. This initialization process could be done quickly if the control computer was fitted with a bar code reader. If not, the converter identification number could be entered on the keyboard. Manufacturers of addressable converters supplied the control software to the cable operators. This software was used to program the controlling computer, which in turn controlled the converters out in the field. During the initialization process, the subscriber's identification code number was assigned and added to the database along with the coded information, as the premium services was also added. The control computer sent converter activation refresh signals along with premium channel selections to the addressable converters on a rotating basis. The more premium subscribers a system had, the longer the refresh time cycle. If a converter was unplugged or disconnected from the cable system, it would automatically shut down and become dormant. This prevented passing the addressable converter around or, as they say, the box from walking. Still, actual signal security depends on the level of signal scrambling or encoding methodology.

Scrambling Descrambling   The more complex the video/audio signal is scrambled, the more secure the signal becomes. Also, the more complex the scrambling technique becomes, the more complex the descrambling method becomes. It should be obvious that the added complexity of the whole process results in a higher cost. The two basic categories for signal scrambling are analog and digital. Analog methods are often regarded as soft and digital as hard, referring to the signal security level.

As discussed previously, signal-trapping methods using either positive or negative traps are regarded as soft signal security. This signal could be simply defeated by removing the negative trap (usually at the tap) or replacing it with a look-alike dummy trap. The positive trap, which was needed to remove the interfering carrier, could be placed at the subscriber's TV set. Such traps are easy to obtain and can often be found advertised in the rear merchandising pages of electronic hobby magazines.

A more secure analog method is the sync suppression method, which can be done at R.F. or on the baseband video signal. The horizontal sync pulses are essentially suppressed a sufficient amount, causing the television set to lose its horizontal hold. Both methods use a 15.73-KHz sine wave signal to suppress the sync pulse, which is shown in Figure 5-16. In order to unscramble the signal, the reverse process is used. Essentially, a sine wave at the same frequency is used to restore the synchronizing pulse and the brightness level. This sine wave is transmitted on the cable system either just above the FM broadcast band on any FM carrier or is transmitted as

Figure 5-16

Two forms of sine wave horizontal sync suppression methods

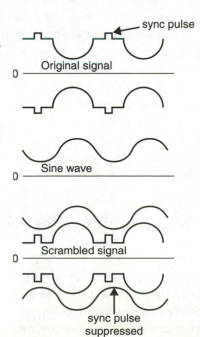

Waveforms for RF sine wave scrambling:

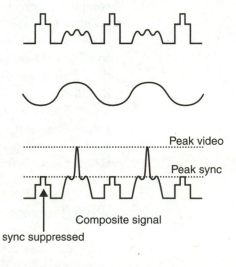

Waveforms for Baseband sine wave scrambling:

part of the FM audio signal on the scrambled channel. The unscrambling process is performed in the set-top converter usually just after the input converter circuit. The restored or unscrambled picture is then converted to the set-top output frequency, usually channel 3 or 4.

Another analog horizontal sync suppression method is the gated sync suppression method. This method can be used for R.F., I.F., or baseband video. In the R.F. and I.F. methods, pulses with the same timing as the horizontal sync for the channel to be scrambled are applied to the RF or IF signal suppressing the sync pulse, usually at six to 10 dB.

Some signal scramblers have the amount of suppression selectable at 0 (no suppression) 3, 6, or 10 dB. Suppressing the horizontal synchronizing signal fools the television receivers sync separator circuits, thus preventing the set to lock the picture.

Other methods provide increased signal security because they further change the video signal. A form of video inversion can be successfully used in conjunction with the gated sync method. Inverting the video signal (between horizontal sync pulses) interchanges the white and dark portion of the picture, resulting in an image like a film negative. Unscrambling is more complicated and hence more expensive to cable operators who feel the higher level of security is worth it. The Zenith Corporation used this type of signal scrambling on their Z-TAC system.

Since digital television is nearly upon us, digital encoding with its highly complex encrypting methods will offer cable operators extremely high levels of signal security. Two methods of digital encoding are line shuffling and line dicing. It should be obvious that at the mention of lines we are referring to NTSC video lines of digitized video signals. By now, one should realize that digital television will not be simply digitized NTSC video signals, but it will be digital data generated by digital television cameras and equipment. Therefore, the security coded algorithms that will be encoded at the point of transmission and decoded within the digital television receiver are yet to be developed. Signal security will also be a high priority that will protect the signals from unauthorized viewing.

The Interactive Subscriber Terminal

The subscriber modem as the set-top box will be a smart converter and will be addressed by the cable system head-end controller, which will respond with an acknowledgement. In essence, the head-end controller will correspond with each subscriber's cable modem using digital handshaking protocols. Cable Labs have recommended that the cable television industry adopt the open cable standard. This specifies a standard set-top unit that subscribers can purchase at local electronic outlets. A conditional access device will be available from the cable operator and will control access to the premium program offerings. The set-top and the conditional access device will require both software and hardware components. As yet, actual units have not been delivered to date. Cable systems that are converting some program channels to digital are going to need this type of system. Digital television channels will be available from video discs and in the near future from HDTV. The major vendors of equipment such as *General Instrument* (GI), *Scientific Atlanta* (SA), and Phillips will have digital set-top units available. Signal security will also be digital with many possible algorithms for a very high level of signal security.

Terminal Initialization   Some initialization procedure will, of course, be necessary and may vary among manufacturers. Digital set-tops will contain memory chips and *reduced instruction set chips* (RISC) to perform the error checking and the required signal processing and signal authorization functions. Ideally, the subscriber will purchase the converter of choice at a local electronics store and call the local cable operator to order initialization. The data stream channel will initialize the converter, programming it to provide service on the desired channels ordered. Most cable systems gearing up for digital television service will also be activating the upstream channels and digital communications between the head-end and the converter will take place. Some systems may elect to use a telephone modem in place of the upstream cable service. The use of software information to control and initialize the subscriber converter is similar to that of activating a cell phone.

Encoding/decoding   The channel encoding/decoding work will be done by the subscriber converter, as always has been the case in the past. However, the process will be done in the digital domain since the digital converter will have the smart chips programmed by the initialization process. Interchanging packets of data by some algorithm will encode the video and audio data stream. The decode procedure will have the preprogrammed recipe to reshuffle the packets into the correct sequence. Digital encoding and decoding offer the cable operator the highest level of signal security.

Program Ordering/Pay/view   Digital television service over cable should produce extremely clear pictures and CD-quality multichannel sound. Digital television is being used in the *direct satellite system* (DSS) where decoding is done in the receiver. The service is ordered at the time of purchase and after installation and activation the authorization codes and decode algorithms are sent on the data channel from the up/down satellite link. Essentially, digital signals will be done in the same manner on cable systems. Digital set-tops will operate by sending the authorized and decoded digital data stream to the digital television set. In the mean time, the set-top will convert the data stream to a NTSC-type television signal required by present-day television sets. This process is similar to that done by the DSS satellite receivers. For a premium programming service, a conditional access module containing the decode algorithm will either be a plug-in chip module or a software module download. In the future, the television receiver will most likely contain all of the decode circuitry and a set-top converter/modem will be unnecessary.

Cable Plant Testing and Maintenance

Instruments and Measurements

The maintenance of a cable system is extremely important and necessary to provide quality service to subscribers. Adding two-way services for data communications requires constant monitoring to prevent noise and signal ingress. As long as the FCC holds cable systems responsible for RF signal leakage, cable operators will have to work hard to maintain a tight cable plant. This is also important in reducing reflections and keeping the signals relatively noise-free. In order to perform the required maintenance procedures, calibrated test equipment in sufficient quantities should be available to the technical staff.

Signal Level Meters

The most common piece of test equipment used by cable technicians is the signal-lever meter. This instrument is used to read the magnitude of an RF carrier. Early *signal-lever meters* (SLM) were essentially tunable voltmeters with electromechanical meters indicating the level in micro-volts. Since converting the voltage value to dBmV was more appropriate for cable technicians, the meters were calibrated to indicate the level in dBmV. The detector used by television signal-lever meters is the peak detector and is appropriate for measuring peak carrier amplitude that occurs at the horizontal sync pulse tips. A basic dual conversion signal-lever meter is shown in Figure 6-1. As shown in the figure, the circuit is basically a dual conversion receiver with a peak detector driving either an analog meter movement or digital LCD display screen. Early meters used the standard analog meter movement. The bandwidth of such a meter was quite narrow and on the order of 400 KHz.

Early Signal Level Meters   Throughout the years, the signal-lever meter has gone through many improvements in level accuracy, bandwidth, tuning stability, and in display technology. From an operational standpoint, since

Figure 6-1
Basic dual conversion signal level meter

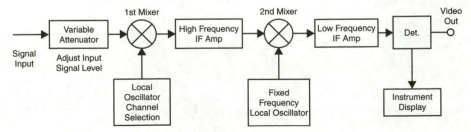

these instruments are portable, battery life has been extended as well as a reduction in weight. Years ago, climbing a pole and carrying a signal-lever meter was hard work. Present-day meters produced by such manufacturers as Sencore, Wavetek, and Sadelco, to name a few, are a bit larger than a hand-held digital multimeter. The accuracy and usability of present-day signal-lever meters are excellent.

Early meters employed a tuning knob and a dial to select the channel to be measured. Also, a switched attenuator was used to adjust the signal level near the center of the meter movement. This attenuator was usually a step attenuator in a number of dB-per-switch positions. Also some instruments needed to be adjusted for frequency response variations for selected frequency bands. Present-day meters using chip technology have extremely flat responses and no band calibrating is necessary. Most meters used today have rechargeable batteries that can easily be charged overnight.

Installer Instruments Most cable systems operating today equip their installation crews with a signal-lever meter to check the tap level at points across the band and at the subscriber's terminal inside the home. This information should be added to the subscriber's file every time it is made. The first time, of course, is at the initial installation and should be measured and recorded at each trouble call. This information can be valuable in tracking system problem areas.

Other test equipment the installer should have is a good, rugged *digital multimeter* (DMM), which can be used to measure the cable system power and the leakage voltage from the ground or between the subscriber's equipment. Cable systems offering telephone service will have to carry the necessary power for the telephone set on the cable system. Such voltage can be measured by the DMM.

A cable system leakage detector should be carried by the installation crew as part of the ongoing signal-monitoring program. Documentation of the tests performed by the installation and service crews can be used to report leakage information.

Proper tools are a must for installation crews. Such tools include the cable connector preparation tools, such as stripping and crimp tools, which are necessary to install leakage and weatherproof connectors. Correct installation procedures can save many future service calls.

Possibly another relative and inexpensive piece of equipment that can provide valuable information is a portable GPS receiver. Subscriber addresses will have a GPS location entered in the plant database. This receiver should have sufficient accuracy (operate in differential modes) in order to differentiate between close subscriber addresses. Software of a table look-up type can be useful in doing leakage location studies as well as installer/ service crew locations using the GPS information.

Multi Function Signal Level Meters   Many present-day signal-level meters can perform several valuable functions. Such functions include determining the signal level of television video/audio carriers, *signal-to-noise* (S/N) ratios, and hum and low-frequency disturbance measurements.

Meters made by several manufacturers use a screen-type LCD display. Displays of this type are driven by digital circuitry, making it simple to store screen displays in digital memory. Periodically, the memory can be loaded into a PC for analysis and storage. Automated test equipment operating in this manner enables the technical department of a cable company to keep accurate records of plant measurements. Plant test records stored on disks or CD-ROMs can be retrieved quickly for analysis and also have the added benefit of not taking up much space. Many signal-lever meters with screen-type LCD displays approach the capability of a spectrum analyzer and are often referred to as the poor man's spectrum analyzer. An example of a screen layout is shown in Figure 6-2.

Spectrum Analyzers

For RF signal measurements, there is nothing quite like a modern spectrum analyzer. Like the signal-lever meter, spectrum analyzers have gone through several major improvements. Not only have they improved in accu-

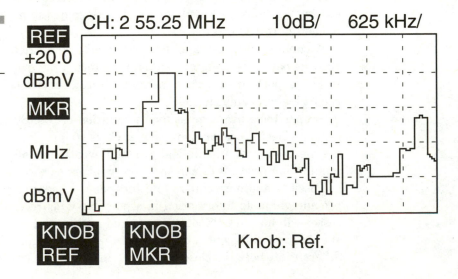

Figure 6-2
LCD screen of TV
CH2 digital display

racy and in its number of features, but several more manufacturers are offering them. Both Hewlett-Packard and Tektronix offer a spectrum analyzer to the cable industry with sufficient test modes and accuracy to perform all the necessary tests required by the FCC for system Proof of Performance tests.

Some of the features that technical personnel consider to be as important as accuracy and performance are simply whether an instrument is lightweight and battery-powered. This makes using equipment in the field a lot more practical. Digital data storage as well as the digital data setup of the instrument is also important to large cable systems that require a lot of measurements. A test setup using such features is shown in Figure 6-3.

Theory of Operation   The spectrum analyzer contains some of the circuitry contained in a signal-lever meter. The display used by spectrum analyzers is still the cathode ray tube, mainly because of its speed and brightness, yet the screen is driven digitally. Some spectrum analyzers offer an analog display mode as well as a digital one, giving the operator the best of both worlds.

A block diagram of an elementary spectrum analyzer is shown in Figure 6-4. This block diagram indicates that the width of the measured signal is essentially the width of the resolution bandwidth filter, which is the last IF amplifier in the spectrum analyzer signal chain. The instrument is also swept through the frequency band at the rate of the cathode ray tube display, following the horizontal power contained in the resolution bandwidth circuit.

Figure 6-3
Automated
measurements and
data logging

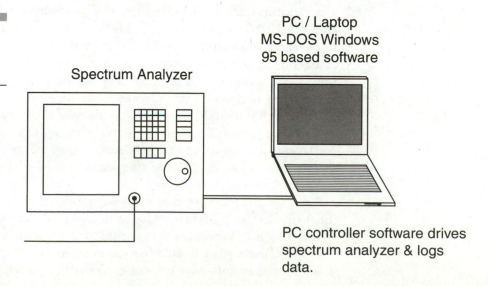

PC / Laptop
MS-DOS Windows
95 based software

Spectrum Analyzer

PC controller software drives
spectrum analyzer & logs
data.

Figure 6-4
Block diagram of
simple spectrum
analyzer

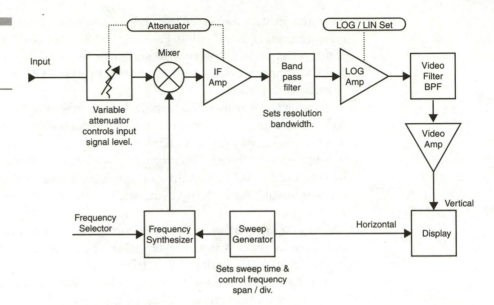

Figure 6-4
Block diagram of
simple spectrum
analyzer

The screen display of a television signal channel is shown in Figure 6-5. In this case, the instrument controls are set to display the entire signal across the six-MHz band. As mentioned earlier, spectrum analyzers on the market today offer many options, and many manufacturers offer analyzers geared to cable television applications.

Spectrum Measurements. Spectrum analyzers with special cable television features are available from several manufacturers, such as *Hewlett-Packard* (HP), *Tektronix* (TEK), and IFR, to name a few. The HP 8590 series has a feature that can personalize the instrument for cable television applications using a plastic magnetic-strip program card inserted through a front-panel access slot. A technician or engineer can program the instrument using a selection of program cards for various applications. This instrument has digital output jacks on the rear panel that can be connected to a PC or a printer. Spectrum analyzer measurements can be printed out on the spot and be placed in a notebook, or they can be saved and dumped to a PC later. When many measurements are needed, such automation features can become a time-saver.

The TEK 2714/15 spectrum analyzer is a made-for-TV type. This instrument can display in either analog or digital mode. It too can have its controls set by a PC and also save digital data measurements. The use of a lap-top PC connected to this instrument can be a big time-saver when many measurements have to be made. Instruments with such time-saving

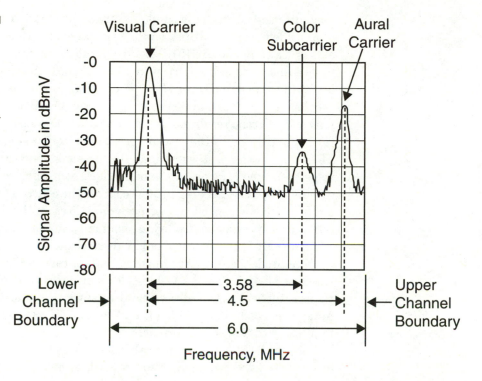

Figure 6-5
*Spectrum of a
television signal
spread across the
whole spetrum
analyzer screen*

features for instrument setup and data cost more, of course, but are well worth the expense, considering the time saved. Systems with data storage and analysis software can save further time and effort in keeping track of system performance.

The controls of many analyzers today use push buttons and what is known as soft-touch controls to set up the instrument to make the prescribed measurements. Many have a data entry keypad to further control the instrument. As might be suspected, it takes quite a bit of time for technicians and engineers to learn all the major features and controls to make accurate signal measurements. Cable Television Proof-of-Performance by Jeff Thomas is applicable to the HP 8590 series spectrum analyzers. TEK also has application notes available using the 2714 spectrum analyzer. It would be a good idea for technical people contemplating the purchase of a spectrum analyzer to send for the application notes and make a study of the various features that will be important to their needs.

Spectrum Analyzer Features & Functions   Once purchased, the next step is to learn to properly operate the instrument and practice making measurements. Basically, any spectrum analyzer measures a signal's amplitude in dBm or dBmV on the vertical axis of the display. The vertical axis is scaled in usually 10 dB or 2 dB per division, and some analyzers also have a linear scale. The horizontal axis indicates the signal's frequency and is usually scaled in KHz or MHz per division. The measurement bandwidth is selectable and should be properly selected based on the speed of the display sweep. If the resolution bandwidth filter does not have time to respond to the signal due to the sweep speed, an error in signal level will result. Therefore, in the past, some manufacturers of spectrum analyzers locked the sweep time controls to the bandwidth control so the problem could be avoided. For those needing to have control, the controls could be unlocked in order to document needed measurements. Present-day spectrum analyzers are micro-processor-controlled to avoid most of these problems.

As a rule of thumb, you should vary the resolution bandwidth if the signal amplitude increases the maximum amplitude settings. This topic is also explained in Jeff Thomas' Cable Television Proof-of-Performance. Information on this subject is also covered in several application notes by TEK. Many cable technicians using the TEK 7L12/7L13 spectrum analyzers are familiar with this characteristic.

Essentially, the spectrum analyzer measures signal amplitude and frequency to a high degree of accuracy. Also, by observing the spaces between signals, noise and spurious signals can be measured. When a spectrum analyzer is placed in the zero-span mode, it acts simply as a tuned receiver and the modulation components can be observed. Essentially, in the zero-span mode, the analyzer acts like a signal-lever meter with an oscilloscope read-out. Now the detected signal can be displayed in the time domain.

For television channels, the signal displayed will be the base-band video signal, with which we are all familiar. The IF bandwidth, or the resolution bandwidth, has to be wide enough to pass the video signal. By displaying the video signal with a slow sweep speed, hum and low-frequency disturbances can be observed. For the hum to be measured, the analyzer has to be placed in the linear vertical scale mode and the operating instructions pertinent to the particular analyzer should be followed. A high-quality spectrum analyzer can be used to make all the measurements required by the FCC Proof of Performance RF tests. Specifications for a spectrum analyzer used for cable television systems are given in Table 6-1.

Other features that are helpful in using the analyzer make accurate measurements are

1. Frequency and amplitude markers

2. Analog CRT display

Table 6-1

Minimum specifications for a Cable System Spectrum Analyzer

Frequency range:	10–1,000 MHz, usually 9 KHz–1,800 MHz
Frequency spans:	0, 100 KHz–1,000 MHz, usually 100 MHz–1 KHz, 1-2-5 steps
Frequency accuracy:	+200 Hz, typical counter accuracy 5 x 10-7
Relative amplitude accuracy:	+2 dB over frequency range, flatness + 2 dB
Maximum input level:	1 watt (damage level), typical +20 dBm max or 68.8 dBmV
Sensitivity:	Minimum signal level -60 dBmV minimum CATV signal
Noise floor:	-60 dBmV with narrowest resolution BW
Internal distortion products:	<60 dBc with analyzer's mixer input 10 dBmV. Various analyzers may state this differently.
Resolution bandwidths:	1 KHz, 3 KHz, 4 MHz video BW, usually analyzers have res BW of 5 MHz, 1 MHz, 300 KHz, 100 KHz, 30 KHz, 10 KHz, 3 KHz, and 1 KHz.
Video bandwidths:	«equal to resolution bandwidths
Input attenuator:	variable 0–60 dB (10 cB steps) some analyzers have multiples steps of 2dB each.
Input preamplifier:	(internal or external) gain > 20 dB and noise Figure <7 dB.
Input impedance:	75 ohm

3. Choice of sweep speeds in zero span mode (10 for full span)

4. TV sync trigger

5. *Fast Fourier Transform* (FFT) circuit on detected output

6. FM demodulator

7. Television picture and sound on CRT display

8. Negative peak detection

Cable & Passive Testing

The signal transport method used in the cable television industry is the coaxial cable itself. This cable is connected together using active devices (amplifiers) and passive devices (taps, couplers, splitters, power inserters).

Testing these devices should be done before installation to assure that no faulty components get installed in the system. A program of testing the devices before installation is termed admittance testing. The test procedures for such equipment are for frequency response and loss.

Cable Loss Measurements Loss at the upper- and lower-frequency bounds for cable and passives are basic and most important. Loss measurements are simple to make and are shown in Figure 6-6. A fixed level of signal is injected and simply measured at the output and should be performed at the lowest and highest frequency. It is known that the cable or device can pass the signal at these two frequencies, the upper and low frequency bound. Many systems make only this measurement for the admittance test. Comparing the loss figures with those given by the manufacturer can give an indication of the length of cable on the reel.

Cable Sweeping Testing A signal sweep test can provide even more information about cable and passive performance characteristics. The results of a sweep test tell the loss of the device over the operating frequency band. Any differences in loss as a function of frequency can be observed. This measured information can be compared against the manufacturer's performance specification that verifies equipment as satisfactory. Figure 6-7 shows a sweep setup for cable testing.

Return Loss Measurements Another test for passive equipment is a return loss test. Return loss in dB is simply equal to 20 times the logarithm of 1/reflection coefficient. This test is performed using a return-loss bridge driven by a signal source and is shown in Figure 6-8. This test basically is a measure of the characteristic impedance variations from the matched condition. Variations in impedance match can cause

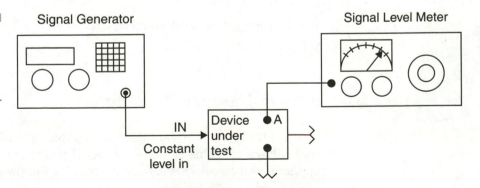

Figure 6-6

Output port A is tested at f_{LO} & f_{HIGH} and level measure by signal level meter

Figure 6-7
Sweep testing of coaxial cable

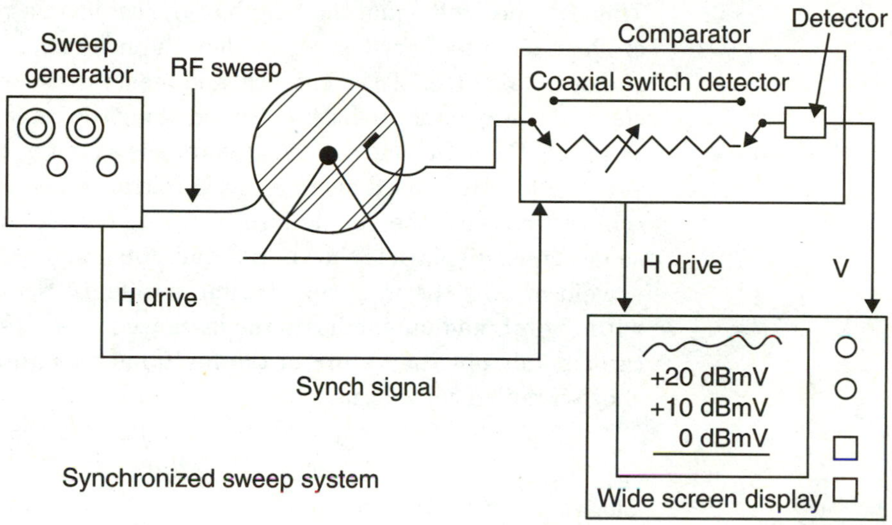

Sweep generator

RF sweep

Comparator

Detector

Coaxial switch detector

H drive

V

H drive

Synch signal

+20 dBmV
+10 dBmV
0 dBmV

Synchronized sweep system

Wide screen display

Figure 6-8
Measurement of cable return loss using a wide band white noise source

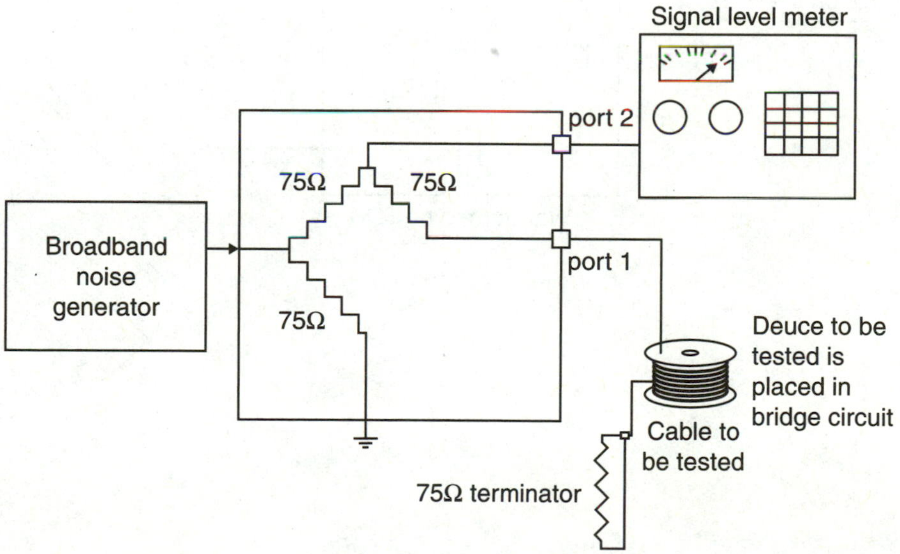

Signal level meter

port 2

75Ω   75Ω

Broadband noise generator

75Ω

port 1

Deuce to be tested is placed in bridge circuit

Cable to be tested

75Ω terminator

signal amplitude variations as well as television signal ghosting in extreme cases.

TDR Testing   Another easy and interesting test to make for cable is using *a time domain reflectometer* (TDR) to measure the length of the cable on a reel. Several TDR instruments are on the market today that operate in the digital domain, having an LCD screen that indicates the distance to a fault.

This distance will equal the length of the cable on a reel if the end is open or short-circuited. For testing the length on a reel, it is usual to leave the end open (a matter of convenience) and measure the distance to the fault (a cut cable). Any abnormalities observed along the cable length will be shown on the LCD display and can be measured as a dent or crush. This TDR measurement is often used alone as an admittance test for cable or is used in conjunction with the simple loss test. The test setup and an example of some screen displays is shown in Figure 6-9 a and b. A TDR displaying the waveform tells the most about cable problems. Some lower-priced TDRs with digital read-outs indicate the distance to the first fault. An LED indicator points out the nature of the fault, such as an open (cut) cable or a short-circuited cable (crushed).

Figure 6-9a
Test for length of cable on a reel.

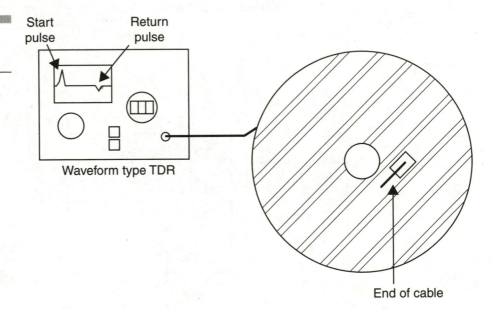

Figure 6-9b
Measurement of distance to cable fault for buried plant.

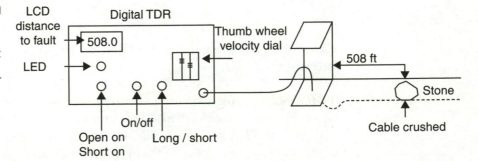

Cable System Tests & Measurements

Cable television systems are tested as mandated by the FCC, which is referred to as a Proof of Performance test. Cable systems are continually tested for cable system signal leakage. Some cable systems that are completely coaxial and are driven by amplifier cascades have a much more extensive problem keeping cable signal leakage under control. Also, the long cascades of trunk amplifiers should be checked periodically for noise and distortion problems.

Cable systems that use fiber-optic technology in which the trunk cable system consists of optical fiber carry both the downstream television signals as well as upstream digital data signals. Such cable systems have a lot less signal leakage problems, since leakage only can exist in the coaxial feeder plant. Also, the coaxial feeder plant usually consists of short RF amplifier cascades. Many systems limit the amplifier cascade to five. These systems also enjoy the fact that cascade noise and distortions are minimal or unobservable. Still, as long as there is coaxial cable plant, RF leakage has to be controlled and must undergo the FCC Proof of Performance testing.

System Turn On & Balancing

The first test any cable system should have is the construction contract compliance test, which is essentially the same as the FCC Proof of Performance test. Passing this test means that the system fulfills the contract specification that it complies to FCC standards.

Initially, at system activation, the pilot frequencies used for amplifier slope and gain are turned on. Next, the amplifiers are adjusted for gain and slope. Other signals can also be injected, but the pilots are necessary. Now the activated plant should be leakage-tested using the pilot carriers. A leakage test transmitter can be placed at the head-end, which has an easily recognized signal that can be detected by portable leakage test receivers. Several manufacturers have such devices available to the industry. Use of these special pieces of equipment can speed up these tests and the cable operator can use this equipment as part of the ongoing leakage control program. Leakage testing done at this time assures that the new system is signal-tight and ready for initial turn-on and balance.

Initial Signal Level Balance Procedure   This initial balance procedure begins at the head-end and progresses down each major trunk leg, balancing each trunk amplifier for slope and gain. At each bridging point, the feeder legs are then set and adjusted so when the end of the trunk leg is reached, all the connected feeders are active. Many systems will connect subscribers as each feeder leg is activated. This helps company image as well as cash flow.

System Sweeping & Response Tests   Once the signal levels are set, the system can be frequency-swept. Companies such as Tektronix and Wavetek have specialized system-sweep testing equipment that provides accurate results and relative ease of use. Many companies often depend on the construction contractor to sweep the system with the work monitored by one of the system's technicians. Some cable operators do not purchase such sweep equipment but may rent or lease it when needed. A system sweep setup is shown in Figure 6-10. Large companies with several systems often have a sweep system that is used on a rotational basis.

Proof of Performance Testing

The FCC Proof of Performance test is mandatory and all systems with coaxial plants have to comply. The commission adopted new regulations in 1992, calling for higher levels of performance. The tests are the same, but the performance standards are raised. Results of the Proof tests have to be carefully documented and placed in the cable operator's technical file. Essentially, the FCC reinstated and revised the Part 76 technical standards pertaining to biannual measurements at widely separated test points in a cable system's

Figure 6-10
Noninterfering
system sweeping

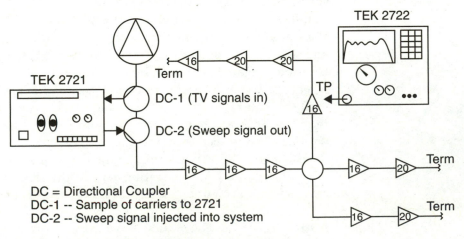

Automatic Sweep System using TEK 3721/2722

distribution plant. One measurement point should be at the longest amplifier cascade. For the first three years starting at February 1992, only RF measurements are required. From then on, demodulated base-band video tests have to be performed on a three-year cycle. The RF measurements consist of signal measurements performed at the test points as well as a system-wide Cumulative Leakage Index of 64.

System Leakage Tests The FCC regards the system leakage tests as extremely important. In the past, it was feared that cable signal leakage occurring in the air safety and navigation frequency bands could constitute a hazard to air safety. The *Federal Aviation Administration* (FAA) pressured the FCC, which in turn pressured cable operators to clean up their plant. With it all, it is extremely doubtful or unheard of that any serious conditions are caused by cable system leakage. However, some cable operators reported fewer plant problems and service calls when the leakage problems were fixed.

Many cable operators have taken advantage of equipment made specifically for finding cable system leaks. Essentially, a transmitter with an easily recognizable audio tone is placed in the head-end and combined with the TV signals. Receivers in the service trucks can monitor this frequency, which aids in finding the source. These receivers are sensitive and can pick up the leakage from the pole plant. Once the area of the leak is found, technical personnel with hand-held receivers can pinpoint the problem. An ongoing leakage control program integrated with the maintenance program means that passing the cumulative leakage test for Proof of Performance is likely to happen. The accepted method for CLI testing, however, is done with a dipole antenna and a signal-lever meter. This procedure appears in Appendix A. The leakage specifications are given in Table 6-2.

Most cable systems use "on-the-ground" testing methods for monitoring and repairing system leaks. However, some large systems covering many areas may elect to do an aerial test. This is usually done by specialized

Table 6-2

Query: Missing Caption. Please Insert

Frequency Band MHz	Field Strength in MV/meter	Test Antenna/distance from Cable Leak in Feet
<54	15	100
*54-216	20	10
>216	15	100

*108–137 is the aeronautical frequency band. It is a good idea to test a carrier in this band for leakage.

contractors who have aircraft outfitted with monitoring equipment and special antennas. This method is known as a fly-over method. Some cable operators feel that this is a more realistic measurement, since the whole leakage problem is concerned with the interference of air safety communications. Equipment in the aircraft can provide data through the measurements to map hot spots of radiated signals. Now the system technical personnel can converge on these areas and repair the actual leaks. Documentation of measurement data, the time and date of measurement, the calibration of instruments, and the calculation of CLI is important information to be filed in the system's technical file.

FCC Test Requirements   The signal test requirements for RF and base-band video signals are stated in detail in Part 76 of the FCC regulations. The RF signal tests are conducted twice a year. Therefore, *multiple system operators* (MSO) usually have a proof of performance crew going from system to system on a rotating basis. Technicians doing this work get good at doing it through the constant practice. The base-band signals also have to be tested.

The FCC rules state that systems with less than 1,000 subscribers are exempt from the Proof of Performance test. Systems with 1,000 to 12,500 subscribers have to perform the test at six widely separated test points in the system and one-third (2) of the locations have to be at points most distant from the system input (head-end or signal source node). One test point is added for each 12,500 subscribers. The results of the Proof of Performance test must be kept in the company files for a period of five years. Records can be either in notebook or file folder form or stored on a floppy disk. If a representative of the FCC calls on a company, the five-year records can then be printed out and presented for study. These records have to include the personnel who made the tests and their qualifications, as well as equipment calibration data. Many system's equipment such as spectrum analyzers, frequency counters, and signal-lever meters must have a calibration check every two years.

Essentially, the FCC regulations require the following basic categories of the testing:

1. Who performs tests and their qualifications.

2. The test equipment used with the last calibration test and the date performed.

3. The signal level at the subscriber's terminal is tested to make sure the signal is at an adequate level to deliver the service.

4. The quality of the signal is tested for frequency, noise, and distortion products. Such testing indicates the system is producing a high-quality signal to the subscriber.

5. System leakage testing shows that the system is complying with controlling leakage below any level that could interfere with over-the-air licensed services.

The personnel performing the tests have to demonstrate competence through school training or correspondence school training certified through a certificate or diploma. The SCTE certification program certifies technician's and engineer's competency in performing the FCC tests. Some cable companies have "on-the-job" training programs that could be acceptable to the FCC.

The equipment used should be periodically calibrated by the equipment manufacturer's service department or by an independent test and calibration laboratory. The date of the calibration service has to appear in the FCC Proof documents.

The testing of a signal level at the subscriber's terminal refers to the input to the subscriber's TV set or the set-top converter output. Some operators perform a test on the input-output characteristics of an arbitrarily selected converter and use this information for this test. Now the signal level of the tap port can be used for a 100 feet of drop cable (3dBmV). Some cable technicians will take a 100-foot piece of cable and connect to a test point tap port. Connecting this cable to a converter in the test vehicle allows this measurement to be accurately made.

The present FCC Proof of Performance tests in force today are for NTSC television signals. When digital transmission arrives, the FCC no doubt will produce new signal tests and specifications for compliance. In the meantime, we will hold our breath. For the present, we will all comply with the present FCC Proof of Performance tests.

It is assumed by now that all cable systems have the latest rules in their possession, and this book discusses helpful tips and techniques. The signal level at the subscriber's terminal can be measured using either a signal-lever meter or a spectrum analyzer. Some systems connect to the set-top converter and then observe the level at the output of the converter, noting any signal-level variations as the converter is tuned through the cable channels. This most likely is not acceptable by the FCC, but signals that are low in levels can be quickly spotted and measured.

Other signal-level tests are to be made over a 24-hour period, which can indicate any signal-level variation between daytime (warm) and nighttime (cool). The FCC specification states the video signal level will not vary more than 8 dB within any six-month interval, which must have four tests in six-hour increments (over 24 hours) during July or August. For January or February, the video carrier level will not vary more than 3 dB within a six-MHz carrier separation, or 10 dB on any other channel. This measurement is performed on the visual carrier of each cable channel. Many systems use three technicians working eight-hour shifts to accomplish this test, while another test crew visits the six or more test points, performing the signal quality tests. Proper documentation of the tests can simply be written on pre-printed forms or stored on a lap-top computer floppy disk. Some computer-oriented systems use a software program to compile and or print the data.

Signal quality tests concern the signal spacing of the video and audio carrier. To do this, usually the visual carrier frequency is measured and then the audio carrier; the difference should be 4.5 MHz + 5 KHz. This measurement can be made by either the TEK 2714/15 or an HP 8590 spectrum analyzer.

Before these new spectrum analyzers were available, a tunable down converter was used with a digital frequency output indicating the visual/audio frequency offset (4.5 MHz). Another manufacturer had a tunable signal stripping analyzer that could read the visual carrier frequency and the audio carrier offset. Such equipment provided the proper accuracy but was more difficult to use and a bit more time-consuming.

Television signals must be tested for noise and low-frequency disturbances (hum). The carrier-to-noise ratio is a particularly important specification. Usually, this appears worse at the longer cascades, because this noise builds up with the number of amplifiers in a cascade. This subject is covered in detail in Appendix F. Since the specification is stated as greater than 36 dB, most cable operators try to keep the *carrier-to-noise* (C/N) ratio at the subscriber's terminal at the longest amplifier cascade to 46 to 47 dB. Subscribers with large-screen television sets find the 36 dB specification allows for the noise to be objectionable. The cable industry should strive to increase the specification to 48 or 49 dB C/N. With this specification, systems can be more competitive with the direct *digital broadcast services*' (DBS) satellite-delivered services. This test can be made with some of the present-day signal-lever meters most cable operators use.

Also, such signal-lever meters can make the hum and low-frequency disturbance tests. This measurement can also be performed using a high-quality spectrum analyzer. When using any spectrum analyzer, a pre-selector filter is needed to protect the input of the analyzer from other signals that can be large enough to overdrive the analyzer's mixer. Strong signals entering the mixer can cause the analyzer to produce spurious signals, rendering the measurement inaccurate. Pre-selector filters are made by Wavetek and Trilithic, to name a couple. Proof of Performance test gear usually has pre-selector (TV channel band pass) filters in a cabinet with each one tuned to the test channels. Such an arrangement of equipment is shown in Figure 6-11.

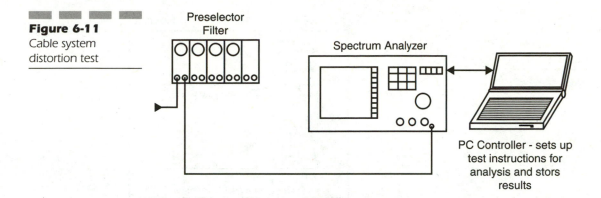

Figure 6-11
Cable system distortion test

System flatness or frequency response tests demonstrate that the distribution system is capable of providing the proper signal level throughout the spectrum or downstream frequency band. A sweep test using a simultaneous sweep transmitter at the head-end and a sweep receiver at the test point is the usual test that provides this information. Systems that are fully loaded can be tested using a spectrum analyzer or one of the new signal-lever meters that display visual and audio carrier levels across the band. Also, the audio carrier level should be at least 13 dB below the visual carrier level, the usual being 15 dB. This prevents sound beats in the picture of the upper video carrier.

In band frequency, the response addresses the frequency signal flatness within the six-MHz channel. The regulation states that the response should not vary +2dB from 0.75 to 5.0 MHz above the lower channel frequency boundary. This test is actually a sweep response test for each channel. Usually, this pertains to modulators and not signal processors. More elegant cable systems that use demod/remod schemes and have the channels phase-locked to a comb generator, providing *incrementally regulated carriers* (IRC) or *harmonically regulated carriers* (HRC), should test the modulators carrying the off-air broadcast television services. A video sweep generator will provide the signal to the modulator and the receiver will detect the RF output response over the six-MHz band. An in-band sweep test is shown in Figure 6-12, including an example of the measurement. This type of test is conducted on the equipment at the head-end. The spectrum analyzer can act as the sweep receiver for the RF. output of the modulator.

If an RF sweep generator provides a narrow RF signal six-MHz wide, then a processor can be sweep tested over the channel bandwidth. Wavetek manufactured the 1855/65 sweep system, which can be used to do all the required system sweep tests. This equipment operates on a battery with an

Figure 6-12
IN band sweep
response test

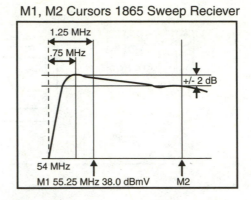

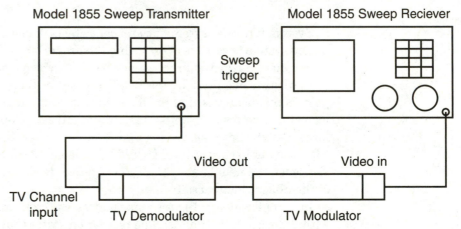

internal charger that provides several hours of operating time. If the service vehicle has a commercial power source in the form of either a rotary converter or a solid-state converter, the sweep receiver can be charging while driving to the next test point.

As stated earlier, a state-of-the-art spectrum analyzer with an accurate frequency counter built in can be used to test the cable system aural (sound) carrier frequencies to the FCC-required accuracy of 4.5 MHz +5 KHz. This is measured at the subscriber terminal. Usually, this is measured at the head-end and should not be affected by the distribution system or the set-top converter. Other specified frequencies that have to be tested are the channels that operate in the aeronautical frequency bands, which are 108–137 MHz and 225–400 MHz. Specifically, cable channels operating in these bands have to be offset from certain aeronautical carrier frequencies by certain offsets. An offset of 12.5 KHz +5 KHz or a 25-KHz offset +5 KHz is required of cable channels operating at power levels of 10-4 watts, which corresponds to -10 dBm or about +39 dBmV. This level is within the level of line extenders or distribution amplifier's output. Therefore, cable systems have to comply with this offset regulation.

The Proof of Performance test logs should be pre-printed with the channel frequencies to be tested also listed, including the offsets, so measured data can be entered in the next column. This test can be and usually is done at the head-end and is usually performed first or by a different test crew. Frequency testing can be done with a signal-stripping tunable frequency counter or preferably one of the new state-of-the-art spectrum analyzers containing an accurate frequency counter. Frequency accuracy for such instruments should be at least four parts in 10-7. As any frequency counter gets older, the time base system deteriorates. This is referred to as aging and is usually given in the instrument specifications. For example, if the instrument aging specification is 1 x 10-7/year and if the instrument is going to be used to measure one GHz (upper frequency limit), then the measured frequency can drift +100 Hz. This result is calculated by multiplying the one GHz by the aging specification. Regardless of the instrument used to make the frequency tests, the operator should study the instrument's operating manual and practice making the measurements before doing the actual Proof of Performance tests.

In the past, the spectrum analyzer has been the instrument of choice for measuring signal distortions. Most of us know that signal distortion breaks down into second- and third-order distortions. Composite second-order distortion, or CSO, is made from the signals being carried on a cable system. All distortion levels have to be 51 dB below the visual carrier level when the system is not phase-locked (IRC, HRC) and 47 dB for phase-locked systems. These measurements should be averaged over time.

Digital storage in the spectrum analyzer can be very helpful. Most of us are very much aware of the calculations of system beat build-up due to the small amount of amplifier non-linearity. Also the build-up is a function of the number of channels carried on the system and the number of amplifiers in the cascade. Present-day systems using fiber-optic trunking to distribution nodes typically carry many channels (approximately 100–110 channels) through short cascades of usually no more than five amplifiers, and hence such distortions are well within specification. Since the cascade factor, as discussed in Appendix F, is 20 log n, n is the number of amplifiers in the cascade, which calculates out for five amplifiers to 14 dB. For amplifiers with a *composite triple beat* (CTB) of 68 dB, then the five amplifier cascade relates to 68-14 or 54 dB, which is three dB better than the allowed 51 dB specification. Therefore, the conclusion can be made that systems rich in optical fiber, where the fiber-optic plants replaced the coaxial cable trunk and the cascade number is reduced to five amplifiers, the *composite second order* (CSO) and CTB specs will be at least 70 dB. The likelihood of seeing any measurable distortion is quite remote. Also, such systems have reduced the number of miles of coaxial cable plant, thus reducing the possibility of cable leakage. However, the coaxial distribution part of the cable system operating at a high-signal level will have to be diligently monitored for ingress and leakage.

System Maintenance Measurements

A system maintenance program is extremely important in providing good service to subscribers. More important is that many cable operators are adding, if they have not already done so, digital services to residential and business subscribers. In order to do so, the cable operator has to have an upstream or return path that is operating noise-free and ingress-free. The procedures vary from system to system, depending on the network topology. Some systems do a market survey first and then activate the reverse plant accordingly.

For all coax systems, activating the reverse system can be a difficult job. Usually, this is best done by reworking a section, progressing back toward the head-end. For subsplit reverse systems, the subscriber drops should have band-reject filters installed to limit signal input to the reverse. The section should then be forward-leak tested to locate any loose connectors where noise and signal ingress can occur. For systems with an optical fiber trunk, the upstream reverse is optically transmitted from the fiber coax node to the head-end. The reverse signals are the "T" subsplit reverse carriers with a digital modulation of QAM-16, for example.

Possibly, many of us may wonder whatever happened to the mid- and high-split systems that were discussed a few years ago. Evidently, if any were actually built, most likely their trunk systems have also been replaced with optical fiber. Such systems had much more bandwidth for upstream applications, but it should be obvious that a well-maintained system would give us fewer problems when activating the reverse system.

End Level Tests   One of the simplest and most valuable tests is a system end-level test. To use such a test program, ground-level test points should be installed by running a drop from a tap connected near the last *line extender* (LE) amplifier at one of the system ends. If the cable system has a skeleton plant diagram, the end points where tests should be performed will be obvious. A sample of such a diagram is shown in Figure 6-13.

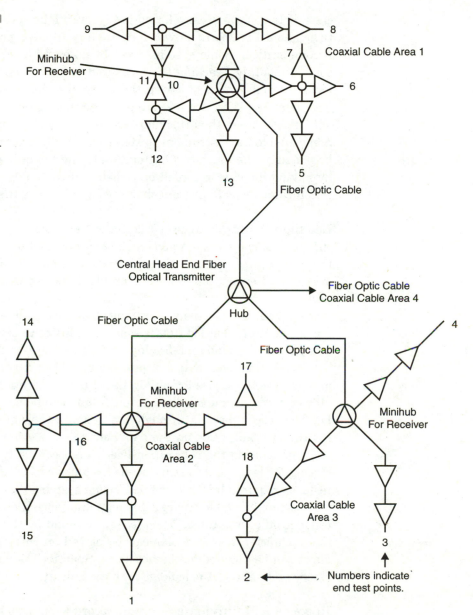

Figure 6-13
Hub-to-minihub fiber-optic AM backbone. Trunk to nodes (minihub) showing end points for end level testing. 18 end test points shown

Systems using long cascades of trunk amplifiers can exhibit poor system flatness responses due to the amplifiers' characteristic signature. A slight dip, for example, in the mid-band can develop to a deep valley down the cascade. In the past, the amplifier manufacturers did not manufacture the amplifiers to tight tolerances, so the dips and small peaks varied differently, causing the effects to cancel. End-level tests on the system's carriers spaced throughout the spectrum will indicate any flatness problems before they become critical. Some systems elect to take end-level tests monthly and some bimonthly. Tracking the measured values through the seasons provides information on how well the thermal compensation and *automatic gain circuits* (AGC) are operating. These measurements are easily made with most state-of-the-art signal-lever meters with a keypad channel entry. Some enable the instrument to step through the programmed channels to be tested by the touch of a button. Analyzing the data can be made easier by having the data recorded on a disk and the analysis software provide graphical plots on histogram charts of signal-level variations.

Leakage Testing Program   The other required maintenance program is that of leakage testing. A good leakage program aids a system in keeping the plant tight, keeping leakage, ingress, and noise to a minimum. Often a system will encounter connector problems before upstream data service problems develop.

Although the FCC requires leakage tests to be made using a test dipole, a more accurate method that is easier and faster to use is the leakage test equipment sets made for finding system leaks. A calibrated leak can be made near the maintenance department and be activated when the install and service trucks are leaving in the morning. White lines indicating the 10-foot distance and the drive-through lane can be painted on the parking lot. The trucks passing through can activate their equipment and check instrument calibration on their way out to work. After leaving, the equipment generating the calibration leak signal can be turned off. Continual testing for leaks and making needed repairs can assure passing the CLI tests, as required by the Proof test. Some systems that have a plant skeletal diagram available enlarge this map and place it on a wall in the maintenance office or garage. When leaks are found by the various install and line technicians, colored pins can be placed on this map. Then the work effort can be directed at the hot spots. Such techniques can increase efficiency and improve the leakage problem quickly.

Records & Maintenance   Proper record-keeping of test results can be a benefit in keeping the cable system in top-notch order. Logs of head-end signal levels on each of the carried television channels can often indicate when a modulator or signal processor is starting to have problems. The head-end signal levels are extremely important for coaxial cable trunk runs. We all should know by now what happens at the beginning of a cascade; it gets multiplied by the number of amplifiers in the cascade. Head-end signal-level stability should be closely monitored and recorded in a head-end log-book or entered digitally using a computer. A computer program that produces bar or histogram figures can be used to plot measured signal-level values for each channel. Such a display can show signal-level variations at a glance.

Some signal-level meters can produce a bar-graph display, and if the instrument has the data available on an output connector, a computer can be used to store the data and/or display it on demand. Most head-ends have a television set or a monitor in the head-end that should be used to examine picture quality for each of the carried channels. A head-end technician should periodically check signal quality at least three times each day. Any deterioration of picture quality can be determined and corrected before subscribers start to complain. Never forget that the picture is most perfect in the head-end. Many systems have several trunk cascades connected to the head-end directly. Each of these trunk cascades should have a head-end test point available and a log kept of each input to the trunk run. Systems with several trunk runs connected directly to the head-end are in an advantageous position when reverse plant activation is to be done.

Some systems have automatic test systems installed that measure and monitor various test parameters and communicate back through an upstream channel to the head-end, giving an alarm as well as measurement data. A cable system may only monitor system power supplies or elect to monitor ending trunk amplifiers only. Online active system monitoring is the most elegant way to go, even though it is expensive.

Head-end Hub Testing

Cable programming comes from three basic sources. One is the television broadcast stations that are traditionally received off the air from tower-mounted antennas. The second is the satellite system cable operator's use of cable-only channels and third is the locally generated cable channels. A fourth is a microwave link still used by some systems.

Off Air Signals

The tower, lighting, and grounding system as well as the antennas and downleads all require a maintenance program. This subject has been well discussed in a variety of sources appearing in the bibliography. Fortunately, some systems have technical personnel on their staff that can climb the tower and perform maintenance work. Periodically, lightbulbs have to be replaced and antenna connections cleaned and remade. Also, the tower grounding should be tested and examined a couple of times a year. Guy wires may have to be tensioned to keep the tower straight. Tower snaking and tilt can be measured using a surveyor's optical transit or can be checked by a local civil engineer. Some cable systems make arrangements with local broadcasters to get the signal by optical fiber. Cable operators with an optical fiber trunk often have spare fibers going right by the television studio facilities. These broadcast television signals will be of studio quality for the cable operator.

Satellite Program Services   Other sources of signals usually consisting of the premium pay channels come from one of the satellite systems. Most cable systems have several parabolic antennas to pick up the various satellite channels for the cable-only channels. In most cases, these channels are scrambled or encoded preventing unauthorized use. Cable systems at the head-end have banks of receivers with built-in decoders called *integrated receiver decoders* (IRD). This equipment provides the cable operator with a base-band video-stereo audio signal that will be re-encoded with the cable operator's equipment and remodulated on a cable channel. The video quality tests are usually done on the base-band NTSC signal, according to the *National Television Committee-Seven* (NTC-7) signal tests. This assures that the NTSC base-band signal is up to broadcast standards. This procedure using a waveform monitor/vector scope appears in many of the references. The grounding of the satellite antennas is extremely important to protect the dish-mounted, sensitive, solid-state *low-noise block-down converters* (LNBC). The mounting frame should be connected to several ground rods around the base by heavy copper wire and bonded to the head-end power ground. Also, keeping snow and debris from collecting in the dish is a must. Pointing this antenna on a satellite is discussed in Appendix G.

Microwave Connections

Some cable systems still use microwave signal sources containing their cable programming. This subject, is discussed in Appendix H. Base-band signal testing should be done on these microwave-received channels to assure broadcast-quality signals are carried on the cable system.

Locally Generated Signals

The locally generated signals are also signals only cable subscribers can receive. In the past, the local origination channels were often considered by cable operators as a burdensome and unnecessary expense. Other cable operators used these channels to carry interesting and informative local programs, thus promoting the cable system. A local studio means that cameras switching both video and audio equipment have to be installed in a proper room. Personnel trained in television program production are necessary to provide quality programming. Again, the picture quality has to be closely monitored and held to NTC-7 standards.

Fiber Optical Plant

Since many systems have or will be installing fiber-optic plants, it is timely to study these methods. A system may not have to install all of its own fiber, since the telephone companies, power companies, and independent common carrier systems have installed huge amounts of optical fiber and may have some fibers available for lease. Fiber-optic theory is available from sources mentioned in the bibliography, but from a practical sense, it will be taken up here.

Essentially, an optical fiber is a waveguide for extremely high frequencies (100 THz) or short-wave lengths (850-1550 nm). These wavelengths are below the visual spectrum. Therefore, we can't see this light at sufficient power levels to cause eye damage. People working around such equipment should wear protective goggles. Also, one has to remember that optical fiber is glass, it can shatter, and it is sharp. Observing proper safety procedures when working with bare or stripped optical fiber is very important. Small pieces of broken optical fiber can get under a person's skin, causing discomfort and possible infection.

Fiber Cable

Fiber-optic systems are characterized as multimode or single mode. Multimode breaks down into either step-index or graded-index optical fiber. Step-index multimode fiber is the first generation, where the cladding (outside) glass has a lower refractive index than the core glass. This refractive index goes from a low to high value in a step change. As discussed in the references, light rays travel in many modes (paths) through the fiber, arriving at the end at different times. Such fibers demonstrate pulse stretching, thus limiting the bit rate. To fix this, manufacturers can produce an optical fiber that has the refractive index to decrease gradually from the core to the outside glass, thus producing graded-index multimode fiber. Now rays of optical energy entering the core can speed up in the lower refractive index of the cladding, arriving at the end more in time with rays that traveled through the core and are slower at the shorter path.

Multimode graded-index became the choice for *local area networks* (LAN) serving digital data communications. In general, multimode fiber is easier to manufacture because it has a larger core and thinner cladding; hence, it is easier to get light energy into a larger diameter core. Lower-cost *light-emitting diodes* (LED) can be used for the transmitter and photo diodes used as the receiver. Multimode fiber systems have limited bandwidth and digital pulse rates. Single-mode fiber is used by cable operators to replace the coaxial trunk system, as discussed in Chapter 4, "Digital Technology and Cable System Applications."

Optical Transmitters Fiber-optic transmitters for single-mode operations are available to cable operators by several manufacturers. General Instrument, Scientific Atlanta, C-COR, to name a few, all offer an excellent selection of equipment. These as well as other manufacturers are familiar with the cable industry and provide the interface equipment needed to go from optical fiber to the RF distribution plant. Usually, cable operators take a group of RF channels and directly modulate a laser-operated optical transmitter. It must be remembered that the laser spreads its power level over a large band, depending on the size of the bandwidth of RF signals. Therefore, for longer distances, the RF band can be split into two or three bands per each transmitter. Then two or more receivers can have their outputs combined. In most instances, high-power transmitters, transmitting through the normal fiber-optic distances, can carry the whole RF programming band. At one time, it was thought that using FM would be better, but technology in AM laser operations improved to the point that it became the method of choice.

Optical Receivers The optical receivers used by the cable television industry is the integrated receiver for the RF converter type. The optical cable is often terminated into the housing and the optical signal is converted to RF at an appropriate signal level, comparative to a normal bridge amplifier level. This signal, as discussed in Chapter 4, is distributed to taps

Figure 6-14
AM fiber-optic
transmission of cable
television carriers

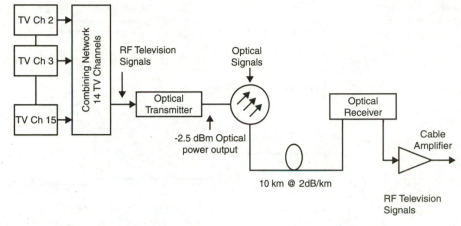

Power Budget:

    10 km cable loss: 20 dB
    Connector and splice loss: 2 dB
    Optical power at Receiver Input: -2.5 dBm - 22 dB = -24.5 dBm
    If the Receiver has a sensitivity of -30 dBm, the margin is:
    -24.5 dBm - (-30 dBm) = 5.5 dB, which is adequate.

by usually a five-amplifier RF cascade. Such an optical system from the head-end to a receiving point is shown in Figure 6-14.

Fiber Optical Testing.

For many technicians, testing optical cable and devices is new and threatening to some. However, it quickly becomes apparent that the optical signal level is measured in terms of power in the quite normal unit of optical dBm. Optical power meters have been developed that are small and easy to use. Also, the optical time domain reflectometer has been developed, which can be used to test optical cable.

Optical Signal Level Testing   Optical signal levels can be measured using an optical power meter that measures the level directly in dBm. Power meters operate at only one or two wavelengths, depending on whether the instrument is a single or dual wavelength type. The instrument range is wide enough so it can be connected via an optical jumper directly to the transmitter. Reading this level as the input level and testing the level at the receiving point can yield the optical path loss. This method

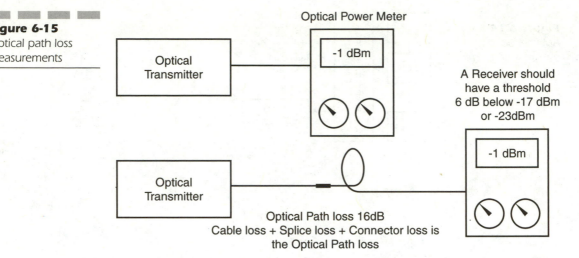

is shown in Figure 6-15. The optical level is displayed on a back-lighted LCD display in numbers of dBm of optical power. The one-milliwatt level corresponds to 0 dBm.

These instruments are hand-held, battery-operated, and usually available with a variety of adapters for different optical connectors. Also, this instrument is quite rugged and not expensive, usually about $500 to $1500. One manufacturer offers a power meter and a light source combination in which a technician can use the light source with voice communications through a spare fiber to an accomplice with the same equipment at the opposite cable end. As a team, the two technicians can test the other fibers in the cable.

Optical TDR The optical TDR is an important and valuable piece of equipment and every system with any optical cable should have at least one. The typical price range for these devices is $6,000 to $16,000, depending on options. These instruments have large back-lighted LCD screens displaying a loss profile of the fiber being tested. Cable systems often use cables carrying many fibers where the various fibers can branch to other parts of the system at splice locations.

Optical cable should be tested before it is installed as well as after it is installed and before activation. The optical TDR can be used to establish the length of cable on a reel by measuring a fiber. The end point can be seen and identified, and any anomalies along the fiber can be seen on the display. Some cable operators do not measure every fiber, but most systems will measure at least one in each buffer tube for large fiber-count cables.

Optical cable can either be a loose-tube or tight buffered-tube construction. Some cable systems choose loose-tube cable for aerial plants and tight buffered-tube for underground plants or when the cable is to be pulled through a conduit. Figure 6-16 shows how some faults are displayed on an optical TDR.

Since the testing of fibers is performed before and after installation, the test data has to be stored or filed for future reference. The *TEK Ranger optical* (TDR) has a keyboard option where cable reel number, cable number, buffer tube identifier (usually a color), and the plastic coating color of the fiber can be entered, followed by the loss measurement. This instrument can store a large number of screen measurements that can be loaded to a computer for storage and/or analysis. An optional disk drive is available as well as an RS 232 Port. Data can be stored and analyzed by Windows-supported software. Most optical TDR instruments are battery-powered and are built for troubleshooting in the field.

Optical Source Testing   Other types of optical test equipment must have a high-level source of optical power that operates in the visible spectrum. Before connecting any optical cables to head-end equipment, the cable can be tested by placing a light source to one end of a cable. Any bends

Figure 6-16
OTDR test
configuration and
traces of problems

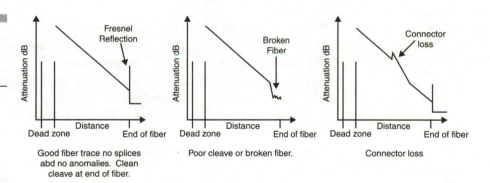

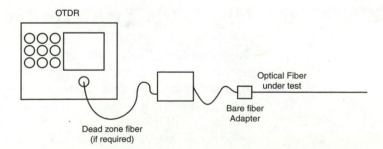

in the optical cable that are too tight or bent too sharp will leak light and can be easily spotted visually. Many systems are concerned that optical cable will be cut, dug up, or damaged and that repairs will have to be made. Single-mode optical fiber is spliced using a fusion splicer with the splices encapulated in sealed PVC enclosures. More and more systems are electing to purchase a fusion splicer and train several technicians in using it. These instruments either are very simple or are microprocessor-controlled and vary in price from $15,000 to $35,000. Also, splicing has to be done in a relatively clean, dust-, and dirt-free environment. Systems usually have a splice trailer or truck outfitted with commercial electrical power, heat, and or air conditioning. The optical cable ends are passed through a hatch door and placed on a brightly- lighted workbench. When the splice has been completed and installed in the closure, it is then passed back out the hatch for system placement. This is quite a procedure that is quite precise but not difficult to learn. Most MSOs and larger systems have such equipment.

Quick fixes using mechanical splices are on the market but are difficult to install and get working. The best method for correcting a cut cable problem is to build in a redundant path for rerouting the optical signal into a node. This type of network topology has to be done during the design phase and built initially into the system. If the signal is lost on the primary fiber, a loss of a signal switch can select the alternate fiber route.

RF. End to End Tests   Since the optical plant is designed to carry the RF television-modulated carriers, testing in the RF domain at the sending or transmission end will establish signal integrity at the fiber cable input. This signal should again be measured at the receiver output for comparison at the input. This type of measurement can be facilitated using one of the present spectrum analyzers that can drive a printer directly. Comparing a print-out of the frequency spectrum of the carriers at each point is a quick way to establish network signal transparency.

Digital Signal Testing

The presence of digital television signals in a cable system again requires more technical training. The broadcast television industry is involved with the conversion to digital signal transmission using 8-VSB modulation. Cable systems plan to use *quadrature amplitude modulation* (QAM) mainly because it will work with consecutive channel operations. Therefore, it is apparent at this writing that cable systems will elect to convert these signals from 8-VSB to QAM-64. Digital signals will have to be tested in the digital domain and in the RF-modulated domain to assure quality transmission.

Instruments for Digital Testing

For many years, digital signals have been used in the telephone and the computer industry. Several instruments have been developed for testing digital signals. The modern oscilloscope with digital sampling, storage, and display is probably the singular most important piece of test gear. The logic analyzer and signature analyzer can test for digital signal errors by recognizing and testing the accumulated errors over a time period. Special test equipment used to analyze specific digital signals for telephone systems has been used for several years to provide the bit error rate information. The cable television industry is also responding to the need for instruments that are needed to test the modulated QAM-64 signals.

Digital Signal Level Meter As mentioned earlier, the signal-level meter is one of the most important and often used instruments in the cable television industry. Signal-level meters needed to test digital modulated carriers are on the market now, with more exotic types just recently introduced. Wavetek, Wandel and Goltermann, Sadelco, and Trilicthic are some of the manufacturers that offer digitally modulated carrier levels. Digital signals in their base-band pulse train conditions can be checked by using a high-quality, high-speed oscilloscope set to produce an "eye" diagram. This diagram is named in this manner because the oscilloscope trace resembles a human eye, as shown in Figure 6-17. Examining this trace can produce information on the quality of the digital signal before modulation. Ideally, the inter-symbol interference should be 0 dB. As most of us know, digital signals can be examined and measured using a high-speed, high-frequency oscilloscope. Digital oscilloscopes offer digital memory and trace storage, as well as wide-band, high-speed operations.

Testing of Digitally Modulated Signals For NTSC television VSB modulation, a negative carrier modulation is specified. This conserves transmitter power, heating, and so on. In short, it is easier on the transmitter system. This type of signal has the peak RF level at the video horizontal sync pulse that occurs every video line for about five microseconds. The detector in the signal-lever meter has to be a peak detector, since this peak signal level is what drives the amplifier cascade to maximum. For

Figure 6-17

Eye diagram of a
digital signal

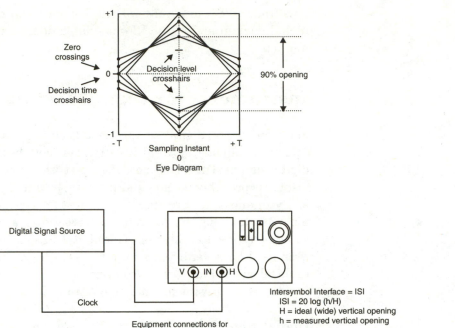

digital signals, an averaging detector is more appropriate. Some of the future signal-level meters will be able to measure NTSC signals with a peak detector and digitally modulated carriers (QAM) with an averaging detector. Present-day signal-level meters offer more options, are hand-held, have a longer battery life, and a larger scanning LCD screen read-out. Some have leakage/ingress options as well as memory for recording measurements. Some of these instruments coming on the market will be able to provide a digital-signal quality check by plotting a constellation diagram of QAM signals. As discussed previously in Chapter 4 and illustrated in Figure 4.32, the constellation diagram is a pattern of dots placed in each of four quadrants. Each dot location in QAM-16 represents a pattern of four dots in each of the four quadrants, for a total of 16 dots. Each dot represents a number in binary code 0000-1111 (0–15 decimal) or four bits per dot. Each dot also represents an amplitude and a phase, hence QAM.

An important new instrument on the digital television scene is made by a relatively new company in instrument manufacturing. This hybrid type of instrument can display a QAM digitally modulated television signal in a variety of modes, such as a six-MHz spectrum (level/amplitude), an average signal level, a histogram display (bar graph), and a constellation diagram in both one quadrant or all four quadrants (Zoom mode). All screen displays are on a well-sized LCD display and are properly annotated. The clarity and tightness of the dots indicate a good signal-to-noise ratio and is related to the bit error rate. This instrument can function in the standard NTSC mode, measuring both the audio and visual carrier levels in dBmV, as shown in Figure 6-18. A digitally plotted spectrum of NTSC television channel 3 is shown in Figure 6-19.

Measuring QAM digitally modulated signals using the Hukk Engineering CR1200 instrument is easy using the simplified keypad and the LCD screen display. QAM digital signal levels can be measured and displayed in a simplified form, as shown in Figure 6-20. Also, the six-MHz-wide band, indicating a QAM 64 signal, is shown as a spectrum in Figure 6-21.

Figure 6-18
Video/Audio carrier level display

Figure 6-19
Digitally plotted display of TV CH3 spectrum

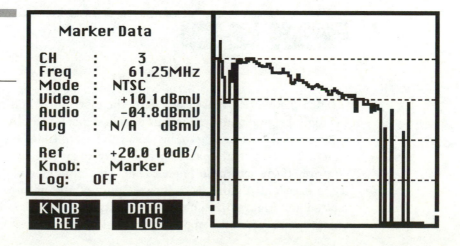

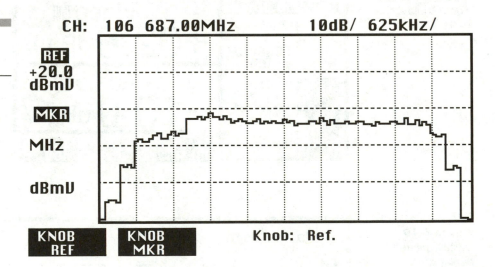

The constellation plot provided by the instrument has a lot to say about
the digital signal quality. The digital signal displayed is an accumulation of
measurements, according to the phase and amplitude positions of the QAM
signal. These measurements are displayed as dots on the constellation
diagram. If the accumulation of the dots is tight and small, the signal quality is excellent. Figure 6-22 shows Channel 109 with all four quadrants
plotted and Figure 6-23 shows essentially the same signal with Zoom activated, showing just the upper right-hand quadrant.

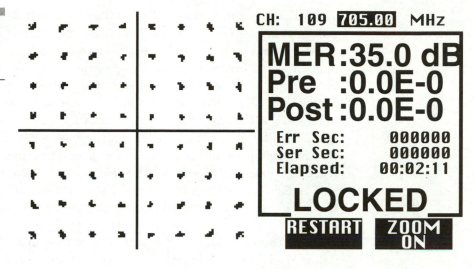

Figure 6-22
Constellation display
of QAM 64 signal on
TV CH109

Figure 6-23
One quadrant of
constellation display
of QAM 64 signal on
TV CH109

The alphanumeric readout indicates MER:35.0dB, which means Modulation Error Ratio and is related to the signal-to-noise ratio. This is measured by comparing dots on the constellation diagram to the position where they are supposed to be. This essentially is a statistical analysis result of the dot location variations. The Pre:3.1.E-3 notation is a measure of uncorrelated data in scientific notation. For example, 3.1E-3 is 3.1 3 10-.3. The post notation is given after *forward error correction* (FEC) has been applied and shows the signal improvement in the same notation as the pre-test.

The MER, Pre, and Post is related to the overall *Bit Error Rate* (BER) but tells a lot more about the signal errors and the causes.

Once operators get used to this type of instrument to measure QAM signals, these features will be greatly appreciated. As of now, only high-speed oscilloscopes and sophisticated spectrum analyzers are available for measuring QAM digital television signals. No doubt, other test equipment manufacturers will offer such equipment, but this company offers the CR 1200 QAM monitor/signal-level meter that comes with a battery-operated, weatherproof case and is reasonably priced. Personnel trained in its use will find it extremely useful in maintaining QAM modulated digital television signals. The screens in Figure 6-22 and 6.23 show QAM 64, the digital format of choice for cable systems offering digital television service.

SONET Testing   Many cable systems operating in the larger urban areas will no doubt find some sources of television programming will be available from optical fiber systems operating with SONET encoding. Now the job at hand is to extract from the *synchronous payload envelope* (SPE) the data packets containing the programming information. Also, SONET terminal equipment will have to be installed in the head-end. Technical personnel will have to be trained in the SONET protocol as well as the testing techniques necessary to test the quality of the signal and point to any trouble spots, should they occur. A discussion of this protocol is covered in sufficient detail in Appendix C along with several testing methods using the HP (CERJAC) SONET test instrument. Various configurations of the instrument to system hook-ups are shown and discussed.

Since this instrument contains both a transmission and a receive system, it can be used in end-to-end testing and daisy-chain test loops, to name a few. Such equipment with its options can be somewhat expensive but will often be necessary for systems connected to SONET facilities. It should be obvious by now that with the interconnections to other communication carriers, many cable television systems will be and are players in an advanced cable telecommunication industry. In the foreseeable future, *fiber to the home* (FTTH) will be a reality with fiber to industry and commercial enterprises occurring first. All telecommunication systems will be carried on optical fiber delivered to and from users. Technicians and engineers will have to get training on optical fiber technology as well as high-speed digital signal-testing methods. System maps, test data, and maintenance records should be stored in digital files for fast retrieval and analysis.

Typical System Problems and Solutions

System Powering and Power Supplies

For many years, coaxial cable systems have carried their own power for operating the cascades of amplifiers. Power supply voltages have increased from 30 volts a.c. to 90 volts a.c. over the years. The most common voltage was 60 volts, which increased to 72–75 volts, and then to 90 volts. Most power supplies used a ferroresonant transformer and was a non-stand-by type. The ferroresonant transformer provided a regulated alternating power supply and a large capacitor was connected across a separate transformer's secondary winding to form the resonant circuit. A circuit diagram of this type of cable power supply is shown in Figure 6-24. Voltage regulation was usually on the order of less than five percent for most supplies of this type. This type of power supply was a workhorse for the industry. Many small or rural systems may still have some in service.

Many systems have replaced such power supplies with stand-by types. Basically, there are two types of stand-by power supplies: the switched supply and the uninterruptible supply. The switched supply turns to stand-by

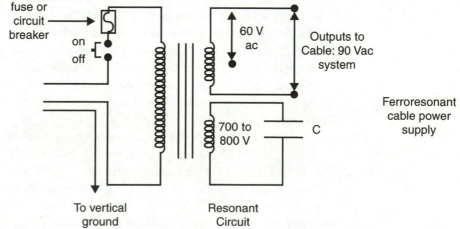

Figure 6-24
Ferroresonant 60-90
v.a.c. power supply

fuse or circuit breaker

on

off

60 V ac

Outputs to Cable: 90 Vac system

Ferroresonant cable power supply

700 to 800 V

C

To vertical ground

Resonant Circuit

battery power upon commercial power failure. When commercial power returns, the unit switches off and places the battery bank on charge.

Early units used relay-type switching where the switching time was on the order of a quarter of a second. Later generation supplies using solid-state switching usually switched in a few milliseconds. Standby times were usually two to four hours, depending on how many batteries were used.

Several models with various stand-by times were available from several manufacturers. Some early supplies used the ferroresonant transformer as a basis. The switch would disconnect the commercial power from the transformer primary and connect a solid-state 60-Hz inverter powered by the battery bank to the transformer primary. It became evident that the core saturation current caused too much battery drain, which decreased stand-by time.

Some manufacturers used two separate transformers, the ferroresonant for commercial power and an ordinary stepdown for the stand-by operation. In stand-by mode, solid-state regulators provided the regulated supply voltage. Unfortunately, these supplies with two transformers and the batteries were extremely heavy and difficult to pole-mount. Thankfully, the uninterruptible supply using a ferroresonant transformer with solid-state regulators and controls is the mainstay of many upgraded/rebuilt cable systems with active upstream and downstream capacity. An efficient design gives stand-by times of up to two hours using two batteries, but most units have three or four batteries for extended on times.

Several present-day designs use modular construction, in which the electronic control and inverter systems can be housed in various cabinet sizes that can contain different battery configurations. Front-panel test points, an LCD read-out of status and test information, and breaker switch and dual outputs are some of the main features of these devices. Problems associated with these power supplies can be devastating to a system, since they were installed to solve the power outage problem and increased signal reliability. Therefore, proper installation and maintenance are extremely important.

As with any pole-mounted equipment, the weather and seasonal changes provide a difficult working environment. Seasonal temperature changes can be as great as 100° F or more with the humidity ranging typically from 40 to 100 percent. Effects of lightning and associated power surges also have to be absorbed by the cable power system. Present-day power supplies used by most of the major systems use 90 volts a.c. to power the plant with a maximum volt ampere rating of 1800 v.a. Some of these supplies offer dual voltages and can be used to supply 60 volts a.c., 72075 volts a.c., or up to 90 volts a.c. Since keeping these power supplies in top-notch condition is extremely important for signal reliability, a full-time, properly trained maintenance crew is required to visit each supply and perform maintenance checks on a periodic basis. Batteries should be checked for any leakage, corrosion at the terminals, and any collection of foreign material collecting in the cabinet. Electrical measurement data for each supply should be entered into the plant maintenance records.

Short and open circuit problems   Problems with cable system powering breakdown into short circuits of open circuits and cause an outage to occur. Troubleshooting the sources of a short or open circuits can be time-consuming, particularly if the area affected is large. Driving back and forth making measurements is the "fencing with windmills" syndrome. Systems using status monitoring on their power supplies usually can solve these problems quickly. Some systems elect to only place status monitoring on the power supplies. Data taken from the power supply's status monitors can be scanned at the maintenance office and the problem can often be pinpointed. Parameters from the status monitor can determine if commercial power is present or whether the supply is in stand-by or normal mode, battery voltages, current draw, or, of course, open or tripped circuit breakers. Usually, the first notification of a problem is usually a signal outage reported by a subscriber by telephone.

The proper methodology is to dispatch a technical repair crew to the general area of the problem while recording calls from the affected area. The repair crews should have available copies of system maps, either on paper or on disk, that can be shown on a lap-top computer monitor in the repair truck. If the affected area is confined to a system power supply, then the repair crew should check the status of the unit. If the supply has an ammeter or an LCD display showing the current drawn from the supply, an excess of a current can indicate a problem. If the indication of the current is changing, then an open or short circuit in one of the system branches could be causing the current variations. Figure 6-25 illustrates this sort of problem.

As shown in the figure, the current provided by the supply flows to the power inserter where it splits in both directions. Ideally, equal currents, as calculated by the design parameters, flow in both directions. In many cases, due to plant additions, this equal current split is destroyed. In order to isolate each current's direction, an ammeter can be clipped across one of the fuses, which can now be removed so the a.c. ammeter can read this current. This procedure is shown in Figure 6-26.

Now the high current indication on the ammeter identifies the cable section with a short-circuit condition. Some manufacturers still offer the self-

Figure 6-25
Power supply current
distribution

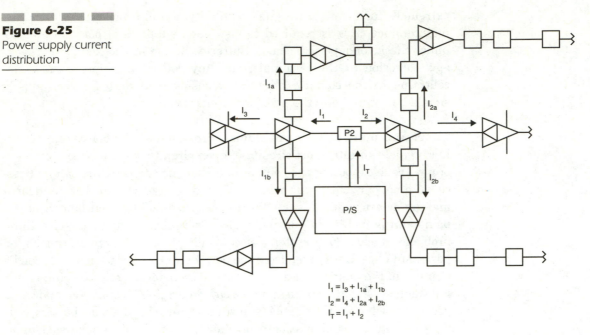

$$I_1 = I_3 + I_{1a} + I_{1b}$$
$$I_2 = I_4 + I_{2a} + I_{2b}$$
$$I_T = I_1 + I_2$$

Figure 6-26
Ammeter positioned
to measure current.
To measure, remove
fuse, read meter and
replace fuse.

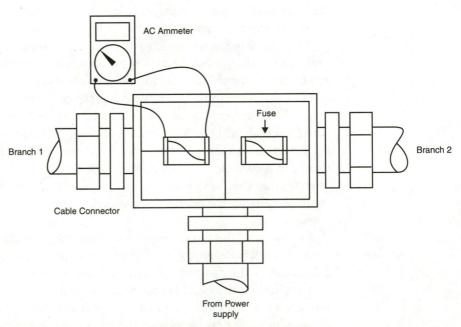

resetting, glass-enclosed circuit breakers, which can be connected through
clip leads across the fuse holder in the power inserter. By observing the con-

tact positions through the glass envelope, a technician can identify the high current leg of the cable system. This method is illustrated in Figure 6-27.

Once the portion of the cable system with either the open or short circuit condition has been identified, the next task is to track down the cause of the problem. Usually, the next area to check out is where next cable split occurs. Either a line-splitter or a directional coupler will be at this location and usually the output ports are fused. Substituting either an a.c. ammeter or a glass envelope circuit breaker will identify the cable section with either an open-circuit (no current) or an overload (high current) condition.

Since power to the feeder or distribution system is derived from the trunk through the bridger, identifying the defective cable section requires some detective work. Trunk amplifier bridger ports are fused and also contain some form of surge suppression. Such surge suppression devices range from gas surge suppressors or the *metal oxide varister* (MOV) types. Examination of the fuse can be done by setting the DMM in a.c. voltage mode and measuring the voltage across the fuse. When the meter indicates full voltage, the fuse is blown, identifying the high current cable section. Conversely, zero voltage, or nearly so, means the fuse is OK and passing current.

Tracking down power distribution problems can be frustrating and time-consuming. In many cases, a cable system plant starts out properly designed and constructed. However, plant additions and adjustments that normally progress in a maturing plant can upset the power distribution balance. This results in different currents in the two parts of the power inserter device.

Figure 6-27
Self resetting
circuit breaker

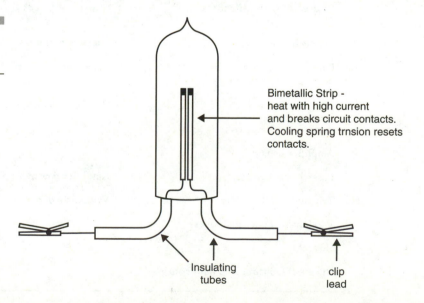

Bimetallic Strip -
heat with high current
and breaks circuit contacts.
Cooling spring trnsion resets
contacts.

Insulating
tubes

clip
lead

Tracking down these open and short circuit conditions in the cable distribution system can be laborious. The causes of open and short circuits are given in Table 6-3.

Solving open- or low-current conditions and short- or high-current condition problems are types of power supply problems that can be aided by a good record of the correct values of the currents. Onsite, current measurements compared to the correct values will point the way to the cause of the problem.

Effects of lighting   The effects of lightning can cause many problems for the power distribution system. Nearby lightning strikes cause large electric and magnetic fields, which induce large voltage spikes into the commercial electric plant. These large electric and magnetic fields induce voltages into any conductor suspended in free space. Since most of these conductors are grounded, the currents caused by induction are passed to the ground. Excess currents can cause the heating of connection points, resulting in visible charring. The lightning damage caused by severe strikes can be inspected and repaired. Power surges in the electric plant can also cause power supply circuit breakers and/or the fuse to blow, which can be protected by surge protectors installed in the amplifier housings. Voltage surges that are not corrected by the power supply surge protectors can cause fuses to blow in the cable direction couplers and signal splitters.

Table 6-3

Causes of Open Circuits	Causes of Short Circuits
Connector pull-out	Connectors (pieces of metal scraps)
Loose seizure screws lightning or power surges	Capacitors, short-circuited due to
Blown fuses	Crushed cables
Corrosion in connectors	
Amplifier housings	
Passive components	Amplifier power supply problems
Tripped circuit breakers	Vandalism of power supplies
Vandalism of power supplies	
Cut cables	
Effects of lightning and power surges	

In some cases, the subscriber taps will become defective after nearby lightning strikes. Some of the ceramic capacitors in the taps will have their working voltage greatly exceeded, due to a voltage surge and burn out. For an aerial plant, taps with a short-circuited capacitor will have a greatly reduced output port signal level. A line technician can remove the seizure screw cover and smell the burned out capacitor, thus identifying the defective tap. This is truly sniffing out trouble. Often, a whole string of taps on a feeder branch can become defective and have to be replaced. Since most taps have a circuit board fastened to the bottom plate, the change-out procedure does not require a splicing technician. Most of the time, inspection of the components on the circuit board will indicate the burned-out capacitor. Most cable systems do not replace the capacitors because other problems with the tap plate may surface later. The distribution amplifier, or line extender amplifier feeding the affected cable run, should be checked for visible damage and for a proper input-output signal level. Unfortunately, lightning problems may take time before surfacing or becoming evident. This can present a lot of difficulty in troubleshooting the affected area and pinpointing the faulty devices.

Stand-by power supplies Stand-by power supplies in cable systems add to the list of maintenance problems. Batteries have come a long way in improvements, however, and several types are available. The battery recommended by the manufacturer is usually the best one to use. Early stand-by power supplies used heavy-duty lead-acid storage batteries similar to those used in cars and trucks. Charging such batteries was critical so as not to overcharge and cause leakage of gases and the electrolyte. Many stand-by units presently available have proper charge controllers as well as solid-state inverters. Still, these units should be periodically inspected and tested.

Stand-by power supply tests are mechanical or electrical. Electrical tests include current and voltage measurements when the unit is in the normal power mode and when it is in the stand-by mode. When in the stand-by mode, the battery voltage should be monitored over a time period of 10 to 15 minutes, which will indicate any significant voltage drop due to battery deterioration. Technicians making these measurements should use a digital multimeter for the a.c. and d.c. voltage measurements. A clip-on ammeter is useful in measuring the a.c. current from the commercial power source and/or the inverter to the ferroresonant transformer.

A visual inspection could determine that loose components have been caused by normal vibration and/or possible vandalism. The most common of the visual problems is battery leakage and corrosion. Such corrosion should be cleaned and any acid or electrolyte should be neutralized with baking soda or an alkali spray. After cleaning, a silicone spray protection should be applied to seal against any action due to moisture. Stand-by power supply inspections should be made on a monthly or bi-monthly basis, and many system operators have a crew of technicians trained specifically for this work. Documentation for these inspections should be either recorded on paper or on a lap-top computer and filed in a power supply maintenance log. A power supply maintenance log form is shown in Figure 6-28.

Probably the most catastrophic power supply problem is that of an automobile striking a utility pole with a stand-by supply mounted on it or when a pedestal containing a stand-by supply is hit. In either case, an outage usually occurs, and if personal injury results, repairs can only be made after the accident scene is cleared. If at all possible, a non-stand-by unit operated by a vehicle generator set can often be jumped into the circuit with clip leads to get the system temporarily back into service. As often said, an "ounce of prevention is worth a pound of cure." Therefore, the installation site for stand-by power supplies should be carefully chosen to be in as safe a place as possible. Corner poles and street intersections should generally be avoided. Pedestal-mounted units should be placed as far off the road as practical and should be marked by a warning stake.

Maintaining the batteries used in stand-by power supplies is a major job. Often, batteries have to be replaced, the result of the periodic testing program. New batteries should be charged and load-tested before deployment

Figure 6-28
Stand-by power
supply log form
example

STANDBY POWER SUPPLY LOG

NO_____ TOWN_____ ST._____ PED POLE#_____

DATE/TECH	TEMP (F)		SEQENCE	AC VOLTS	NORM PWR. LT.	NORM SUB. LT.	INV. RDY. LT.	CHARGE LT.	LOW BAT LT.	BATT VOLTS DC	REMARKS
		NORM									
		STBY									
		8 MIN.									
		15 MIN.									
		NORM									
		NORM									
		STBY									
		8 MIN.									
		15 MIN.									
		NORM									
		NORM									
		STBY									
		8 MIN.									
		15 MIN.									
		NORM									
		NORM									
		STBY									
		8 MIN.									
		15 MIN.									
		NORM									

in the system. The battery maintenance area should be well-ventilated so any out-gassing during charging can be dispersed. Storage batteries or gel cells contain acid as the electrolyte. A jelled acid compound is the electrolyte used in gel cells and can cause serious skin burns if any gets on a person's skin. A supply of baking soda or a solution of baking soda and water should be kept nearby in case of an acid spill.

Also, in the battery room, a charger and a load test should be installed to service any batteries requiring maintenance. Often, one or two stand-by units can be used as a charger since in most instances charging is unnecessary. Switching these supplies into stand-by mode and timing the battery voltage and current can test the capability of the battery to maintain the load. Systems requiring the testing of a large number of batteries may wish to construct a battery charger, as shown in Figure 6-29.

The most common problems occurring in stand-by power supplies involve the battery system. Either the batteries fail or the charger fails to keep them fully charged after an outage. A high internal temperature due to ordinary heating during operations along with a high summer ambient temperature can shorten the life of the batteries. When the power supply is

Figure 6-29
Pulse type battery charger for stand-b batteries

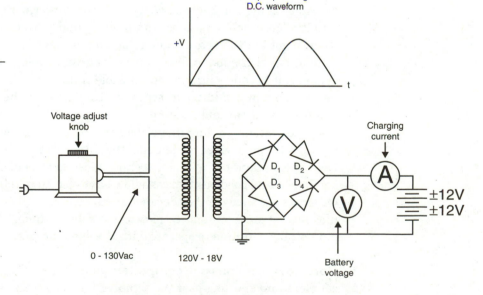

D₁ D₂ D₃ D₄ are 10 amp, 50V diodes
Charger charges 2, 12 volt stand-by power supply batteries in series. Adjust input voltage for desired charging current.

operating in the non-stand-by mode, the ferrotransformer generates heat. Charging the batteries after an outage also causes additional heat, worsening the problem.

Many stand-by power units contain a cooling fan, which can effectively remove heat. Maintenance inspections of stand-by power units should include inspecting the fan system, which may include a thermostatically controlled switch. The internal battery charger should be cycled and checked to make certain the trickle charge current to the charge battery is not too high. This could overcharge the batteries. Most stand-by units have the battery compartment in the bottom shelf of the cabinet, preventing the transformer heat from rising into the battery compartment. As a rule of thumb, the battery-charging parameters, as given by the manufacturer, should be followed when setting the stand-by power units charging controls. In general, stand-by power units that are pole-mounted should be serviced by technicians using a bucket truck.

Coaxial Cable System Problems

The maintenance of coaxial cable system should be familiar to most of us, since we grew up with it. As cable systems age, the cable system trunk and distribution system changes along with the subscriber growth in the community. More cable, passives, amplifiers, and power supplies are added when plant additions are made to accommodate the added areas that are expanding with subdivisions and residential developments. As the plant expands, it becomes hardly recognizable from the original build. Unfortunately, the newer additional areas are connected to the system by the old and problematical cable system.

Cable systems finding themselves in this position, faced with advanced digital television, video on demand, the Internet, and all the additional services, must adapt to these new technologies. Some cable systems that have small debt and good positive cash flow have elected to rebuild their systems. Other systems have either decided to sell out or simply do nothing. Companies buying such systems in most cases rebuilt them. The companies that did nothing continue to operate with the same programs.

Connector and passive component problems   Connectors, directional couplers, and splitters pass the signal as well as system power. Any loose or poorly made connection-conducting current will heat and eventually cause an outage. Connectors, like any other cable system component, have improved over the years.

As stated earlier, the older parts of a system feed the newest and oldest constructed parts of the system. Any problem occurring in the older parts can affect a large number of subscribers. Leakage testing, as required by the FCC, has caused many cable operators to focus on old and defective connectors. A common connector problem is often called a pull-out, in which the cable is essentially pulled out of the connector enough to break the system power. A pull-out can cause power and signal variations, which can be intermittent. Such problems are usually the most evasive and difficult to solve.

Also associated with connectors is the cable itself. Flexing of an aerial plant due to wind, rain, snow, and ice causes strain on the cable, usually at the connector. This strain can cause the aluminum cable sheath to crack open, allowing water and moisture to enter the cable dielectric. Moisture and electricity react to form extreme corrosion and system failure. Most systems use heat-shrink sleeves on the connectors that protect against this. Cable systems in a coastal environment usually have jacketed cable with heat-shrink tubing on the connectors that adds significantly to the life of the cable plant.

Early manufacturers of cable passives paid little attention to the effect different metals had when in contact with each other, particularly with moisture added. Cable systems along the coast often watched their taps and passives literally disintegrate before their eyes. Present-day manufacturers, for the most part, are well aware of the metallurgy involved and make high-quality devices often coated and with protective O-rings and seals.

Construction practices have improved drastically as well. The semicircular or round-bottomed cable expansion loops have been replaced by the flat-bottomed type made by a ratcheting cable bending tool. This type is the most accepted type and is used by most cable systems today. Proper construction practices and methods are discussed in several of the references listed in the bibliography.

System amplifier problems   The failure of amplifiers in the system amplifier cascades can present a myriad of evasive problems. It must never be forgotten that what happens to amplifiers early in the cascade can cause serious problems at the ends. Most systems, at the completion of construction, go through the initial balance followed by sweeping, final balance, and Proof of Performance testing. At this point, the system will never be in better shape. At this point, the so-called shake-down cruise takes place. The leakage testing program usually identifies loose or poorly installed connectors and devices, as well as loose amplifier housings. As a rule of thumb, after the initial turn on and balance, the system should be leakage tested before the sweeping and final balance takes place. This assures that the

final balancing and sweep response testing does not compensate for a leaking and faulty cable plant.

Larger cable systems usually maintain an in-house amplifier laboratory
where an amplifier can be bench-operated, aligned, and adjusted. Amplifiers brought in from the outside plant can be tested to assure that they are
indeed defective before sending them to the manufacturers. Some systems
maintain an inventory of components and integrated circuits, essentially
performing their own amplifier repairs down to the component level, but
many do not elect to change modules and send the defective ones to a contract repair facility. Some systems have an amplifier housing of each type
used in the system setup on a wall with a bench in front where a module
can be tested.

When an amplifier module fails in a plant, a replacement can be pre-
tested and set up, thus speeding the repair process. Such a test setup
enables the technical department to test modules so they can be tagged
with various operating problems before sending them to a repair facility.
This same setup can be used to ascertain that repaired units operate
according to specifications. Large systems often have their own repair facilities that support all their systems in the surrounding area.

Common amplifier problems are often related to the switching power
supply module. This type of power supply is well regulated as per line
power and load power. This type of supply rectifies the 60 or 90 volts a.c.
cable power and then converts this d.c. power to a higher frequency a.c. that
is regulated precisely and then converted down to the required d.c. voltage
used by the amplifiers. The transformer that controls this high frequency
a.c. voltage emits sound at the a.c. frequency. A normal operation sounds
different than an abnormal operation, and often an experienced technician
can tell which is normal. Once suspected, a voltage measurement will often
confirm the faulty unit. Capacitors of the electrolytic type are the most
likely to fail in a power supply. Also, diodes on the input rectifier are known
to fail often after a thunderstorm. Surge suppressors should be inspected
and, if suspected, they should be replaced.

Return system problems   As in the downstream system, the reverse
system can have similar problems, but usually not as many subscribers call
the office. A monitoring system, either manual or a status monitoring system, will indicate a problem and usually identify the general problem area.
In many instances, a technician at the head-end or hub will detect an
abnormal condition on the reverse signal and immediately start to explore
the cause by further testing. Some systems power the reverse system from
the same units feeding the forward. Therefore, failure of a power supply can

affect both forward and return systems. Such a condition will usually point to the faulty power supply.

Since intermittent problems are the most difficult to solve, problems of this nature with the reverse system can at least be monitored at the head-end. The repair crews can go into the system, inject signals, and make adjustments and tests, keeping in touch with the head-end or hub by two-way radio or cellular phone.

Most systems that have an active return system will have test equipment for aligning and troubleshooting the system. As always, a high-quality calibrated spectrum analyzer in the hands of a properly trained technician will usually find the problem unit so a replacement can be quickly made.

Fiber Optical System Problems

Systems using fiber-optic technology usually have another set of problems. Luckily, some of these problems are easier to solve simply because little, if any, electronic equipment is needed between the transmitting and receiving points.

Optical transmitter/receiver problems   Monitoring laser currents at the optical transmitters will indicate laser operations. As lasers age, they tend to draw more current. If the laser in an optical transmitter starts to show an increase in current that is approaching an out-of-normal range, the optical transmitter should be changed out. Still, a laser in the transmitter can suddenly stop operating, causing an outage. If this occurs at the head-end, a change-out can be made quickly, since most systems have a head-end technician on duty most of the time. Laser transmitters on the upstream or return system that fail have to be tracked down and replaced. Optical transmitters and receivers are generally quite rugged and are long-lasting. Most optical electronic equipment contains monitory lamps or meters that indicate normal or faulty operating conditions. Such equipment indicators are extremely helpful in identifying failed or failing equipment.

Present-day optical systems often have equipment test points where an optical power meter or other test equipment can be used to test system performance. Optical power meters can perform field tests for optical performance, but a complete terminal-to-terminal test using an RF spectrum analyzer can confirm system operation. As the optical plant is expanded to *dense wavelength division multiplexing* (DWDM), in which several optical carriers generated by lasers carry many more services, an optical spectrum analyzer can be very helpful. Optical spectrum analyzers essentially display the optical carriers on a CRT screen, similar to that on an RF spectrum analyzer. Such a screen layout is shown in Figure 6-30. The optical carriers can be tested for proper wavelength (or frequency) and optical dBm levels. Lasers not having proper output power can be easily identified and replaced before service is affected. Cable systems that are connected to a SONET link may want to opt for a SONET testing system, as described in Appendix C.

Optical cable problems   Optical cable that is properly installed and spliced together should cause few problems. The most common problem often occurs at a terminal location where the main cable is terminated and connected to equipment using jumpers and pigtail optical cables. Kinking or sharp bends can cause a severe loss to occur. Proper cable tracks and trays should be used to assure that the cable is loosely routed. After installation and before activation, the spliced-in jumper cables can be tested with a light source operating in the visual spectrum and placed at the optical connector. Light will shine through the plastic jacket of the jumper cable if the bend is too tight and can be easily seen and corrected. Once this is

Figure 6-30
Optical spectrum
analyzer display

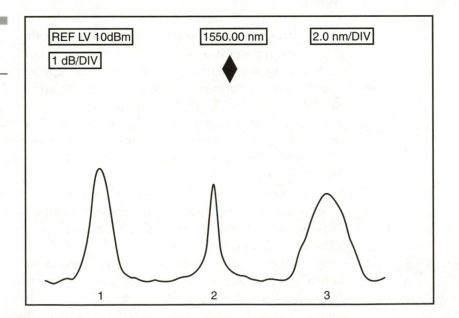

3 Optical signals: signal 1 - +5 dBm, 1540 nm
signal 2 - +4.3 dBm, 1548 nm
signal 3 - +4 dBm, 1556 nm
signal 3 has a wider spectrum width.

accomplished, technicians working at or around these optical jumper cables should be careful not to cause any sharp bends in the cables.

Other problems with optical cables are often man-made. Cars and trucks hit poles, aboveground pedestals, and terminal units, thus causing smashed or broken cables and equipment. A rapid change-out is the only cure for such problems. Operating cable systems have also reported such problems as bullet holes in optical as well as coaxial cable plants. Fires in buildings next to aerial cable plants can damage cable plants as well as lead to forest fires. The possibility of such problems should be addressed at the time of the initial build, but changes in the plant due to extensions and system rerouting can lead to critical equipment being poorly placed. As mentioned earlier, power supplies should not be placed at street intersections. Neither should an optical signal-mode pedestal be placed at a street intersection.

Most other problems with optical plants include keeping it clean and free of moisture. The most common point to remember is to clean the optical connector before reconnecting it to a piece of equipment. Proper cleaning of optical connectors is most important. An almost unseen (even with a magnifier) speck of dust can cause a significant decrease in signals.

Head-end Problems and Maintenance

Head-end problems can be, at times, elusive and confusing, particularly with the off-air television broadcast stations. Some cable systems use a microwave facility to connect remote programming to the hub-head-end facility. In many instances, satellite earth-receiving stations provide a multitude of channel selections, making up the system offering. Often, this installation is done at the head-end site. Each one of these signal sources seems to have some unique problems that have to be solved by the technical staff.

Off-air antenna system problems Local television broadcast stations are received by tower-mounted antennas, appropriately aimed at the station location. The size and position on the tower depends on the required signal strength for the television station signal. Careful study and planning determine the proper antenna and any pre-amplification necessary to provide a high-quality television signal to the head-end. Assuming that the initial design has been adequately done and the system has been properly installed, typical and not-so-typical problems will be investigated. It must not be forgotten that what has been constructed is a tower-mounted listening post where the antenna gain and direction are peaked to the desired station's frequency. What happens between this facility and the station is beyond the control of the cable operation. A problem involving a head-end facility is given in a case study below:

A cable system head-end detects an interfering signal, nearly obliterating a usually good channel 7 broadcast station and rendering it unwatchable. Since afternoon soap operas appearing on this station are coming on, telephone calls from subscribers start to come in. A technician at the head-end is alerted and switches to an auxiliary search antenna. The signal is equally bad on the search antenna and is made worse slightly with a small change in direction. A spectrum analyzer determines that the interfering signal is also a television signal by looking at the detected video.

The chief technician outfits a truck normally used for doing leakage detection and sets the dipole to channel 7, connecting a television set and a signal-lever meter to the antenna. The driver starts traveling toward the source of the interference, thinking that it might be caused by a severe cable leak. Since the signal is so strong, it seems that if a cable leak caused it, other channels could be affected, and they are not. After circling around within a one-mile radius from the head-end, a strong co-channel 7 picture is received on the TV set and a +20-dBmV signal is recorded. In the immediate area is a residential home with a television antenna with the high-band flat lead broken from the high-band antenna section that is drooping down past the antenna mounted pre-amp. This length of twin lead (input to pre-amp) is drooping alongside the pre-amp down lead (output or pre-amp), causing a feed back loop. The flapping lead picks up enough channel 7 signal to produce a strong signal through the pre-amp oscillation. Knocking on the door produces nothing. A neighbor says that the person moved out a while ago. A pair of side cutting pliers is applied to the down lead, thus removing the pre-amp power, and the problem is immediately resolved.

The case just described is a case beyond the control of a cable operator. Clearly, this is not a case of malfunctioning cable equipment or a problem with the cable system, but it resulted in destroying one of the off-air stations. Since most people in the community were cable subscribers and removed their antennas, this problem did not affect them. Troubleshooting, in addition to technical expertise, requires good detective work.

Other head-end problems usually concern failed processors or modulators due to small contact corrosion problems. Cleaning with a contact cleaner often restores operation. Most head-end equipment is so overdesigned, units operate for years going untouched.

Tower-mounted equipment, on the other hand, usually does have failures. Antennas with the wiring harness ty-wrapped are known to have problems. First, some ty-wraps break allow the harness to hang down and swing in the wind. This causes stress at the connection point, allowing the lead to break from the antenna. Clearly, this is a problem that, if noticed, can be fixed. Usually, the tower has to be climbed and the wire secured before it breaks. If only one element is bad, it can be lowered to the ground, repaired, and put back up. Pre-amps and tower-mounted down converters after several years of service also breakdown and fail. In most cases, connector corrosion is the problem. Since the down leads carry power up to the device, the combination of current and moisture causes the failure. Connectors should be sealed with heat-shrink tubing and then be taped.

Microwave system considerations Some systems today import or deliver programming from hub to node using a microwave link. There are basically two types of microwave systems. One is the familiar *amplitude modulated link* (AML) system and the second is the single-channel up-down type.

The AML system operates in the frequency band of 12.2 to 13.2 GHz, known as CARS band frequencies. The up-down conversion type is a short haul system, where the desired band is up-converted to a microwave channel at the transmitting end and down-converted back to the normal television band at the receiving end. One manufacturer provides a system that up-converts the whole cable television band to microwave and transmits it to a receiver that then converts the band back to its normal channels.

Once a microwave system is placed in a normal operation, periodic checks on the transmitter and receiver should be made as a normal maintenance procedure. As klystron tubes or traveling wave tubes age, they deteriorate. Normal monitoring of the drive signals and power supply parameters will indicate the condition of the active elements. If a component is operating on the edge of normal electrical parameters, the component can be changed-out before and break down in service results. Points to inspect are the down-lead cables for any pinching or crushing, antenna pointing brackets, grounding wires, and general corrosion.

Typically, a well-designed and installed microwave system will give many years of service, provided normal maintenance and replacement of components is performed. A common microwave facility outage occurs when the commercial power fails and the stand-by unit fails to work. This problem can be avoided with periodic cycle testing of the back-up power system. Most other problems with a microwave radio link are due to weather damage such as flying objects, severe wind, rain, snow, and ice. Trees or the construction of tall buildings blocking the line of site path is one area beyond the control of the cable operator. The only cure is to relocate the transmit-receive sites to get a clear line of sight path. Replacing the microwave facility with a fiber-optic system is a possible solution.

Satellite earth receiving station problems   Most cable systems have a number of satellite-receiving stations necessary to provide programming from a number of satellites. Breakdowns in any one of the systems can cause a loss of many satellite channels. Getting the system back up and running is of utmost importance.

Some typical problems that are most common are snow, ice, or debris falling into the dish antenna, corroded and damaged cable or connectors, and the effects of lightning. Snow and/or ice build-up in the parabolic dish antenna occurs in antennas with a large elevation angle. Some cable operators use a heating system commercially available from several manufacturers that allows the ice and snow to melt and drain away. Debris such as trees, branches, and limbs can fall into the antenna and often damage the feed horn system. Systems that have problems such as this are not visited periodically by competent technicians who observe the antenna system and the surrounding area. The satellite-receiving site should be visited at least twice a week and the electronic equipment areas inspected as well. In general, a properly installed and maintained satellite system will give years of service.

Some systems still use pressurized, hollow coaxial feed lines, using what is known as heliax. An air pump with a canister of a drying agent is used to pump dry air into the heliax. Breakdown of this pump is when the desiccant turns pink or when the signal from the antenna gets attenuated. When it gets too small, the satellite channel gets sparkles. This again is the result of poor maintenance.

Again, interference from other microwave facilities operating at the same frequencies can cause problems with signal degradation. Some sites may pass on site measurement at the initial construction time, because land clearing and/or nearby construction which eliminates the natural shielding effects can result in interference. The cable operator then has two choices: one to relocate the site, or two to construct an artificial shield. If the situation is not too severe, an antenna shroud or protective shield will effectively reduce side lobes of the antenna-receive pattern to lessen the interfering signal level. Such situations are beyond the control of the cable operator but must solve the problem.

Summary   As the saying goes, an ounce of prevention is worth a pound of cure. Proper location and site planning can and often does avoid potential problems. Placing earth-receiving antennas near trees and structures that can fall on them or cause excessive water, ice, and snow to collect on them can affect an operation. Placing off-air antenna towers or microwave facilities on a hilltop or joining one to a water tank can be a potential problem if any adjoining structure falls or bursts. Setting up facilities, such as an off-air antenna tower, near high-voltage transmission lines can result in electrical interference on the low VHF channels as well as lead to a hazard during bad weather or power line maintenance. Also, locating the receiving antenna tower near AM, FM, or TV-transmitting facilities can present a whole set of problems for the cable operator.

APPENDIX A

Coaxial Cable Leakage Testing and Calculations

The leakage specification for cable television systems depends on the frequency band of interest, the leakage intensity in microvolts per meter, and the measuring distance from the cable plant as summarized in the table below:

Frequency Band MHz	Leakage Level Microvolt/Meter	Distance from Plant Feet
Up to and including 54 MHz	15	100
Over 54 MHz, up to and including 216 MHz	20	10
Over 216 MHz	15	100

The test antenna is specified as a dipole. However, a whip antenna mounted on a roving vehicle is often used to locate leaks, which are then measured with the dipole. The detecting receiver is either a signal-level meter of sufficient sensitivity or a special leakage-detecting receiver. Several instrument manufacturers make a combination test transmitter/receiver pair, and cable operators often have several receivers for ongoing leakage monitoring.

Example: If TV channel 6 is used as a test frequency and the leakage measured is at the 20-μV limit, the following formula can then relate this level to a voltage value:

$$E = 0.0207[F(MHz)]\ [Vr(\mu V)]$$
$$\mu V/M$$

where $E = 20\ \mu V/M$

$$20 = 0.0207 \, [83.25 \text{ MHz}] \, (Vr(\mu V)$$

$$V_{R(\mu V)} = \frac{20}{0.0207 \, [833.25]} = 11.62 \, \mu V$$

This voltage value in dBmV = $-20 \log \dfrac{1000 \, \mu V}{11.62 \, \mu V} = -20 \log 86.06 = -38.7 \text{ dBmV}$

If the signal level meter cannot accurately measure signals a small calibrated preamplifier can be used.

The dipole antenna should be tuned to the test frequency using the following method:

1 wavelength (ft) = $\dfrac{984}{f_{(MHz)}}$ and dipole = $\dfrac{1 \text{ wavelength}}{2}$

Each rod of the dipole will be ¼ wavelength.

Each rod length in feet = $\dfrac{984}{4f_{MHz}} = \dfrac{246}{83.25} = 2.95 \text{ ft}$

and in inches $2.95 \times 12 = 35.4" = 35\frac{1}{2}"$

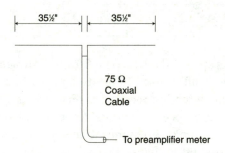

Now that the measured leak level in dBmV can be related to the specified leak in microvolts per meter, a sample calculation of CLI will follow. The method of testing is illustrated below:

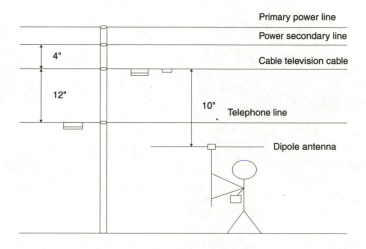

An example of recorded data at TV Ch. 6:

Leak dBmV	Leak mV/m	Leak^2
−39	20	400
−35	30.5	930
−30	54.5	2970
−20	172	29,584
−15	306	93,636
−35	30.5	930
−40	17.2	296
−35	30.6	930
−32	43.2	1866
−30	54.5	2970
	Total	134,512

At a measure level of −30 dBmV corrected by taking any preamplifier gain into account

$$E \text{ mV/m} = 0.0207(83.25)(Vr(\mu V))$$

−30 dBmV corresponds to

$$-30 \text{ dBmV} = 20 \log \frac{VmV}{1000 \ \mu V}$$

$$1.5 = \log \frac{V \mu V}{1000 \ \mu V}$$

$$\frac{V \mu V}{1000 \ \mu V} = \log^{-1} - 1.5 = 0.0316$$

Now: suppose this cable system has 60 miles of total plant and the above data was taken from a 20-mile section. The CLI can now be calculated.

$$V \mu V = 31.6 \ \mu V$$

$$EmV/M = (0.0207)(83.25)(31.6)$$
$$= 1.72 \times 31.6 + 54.5 \ \mu V/m.$$

$$CLZ = 10 \log \frac{60}{20} \times 134{,}512$$

$$= 10 \log [3 \times 134{,}512] = 10 \log [403{,}536] = 10 [5.61] = 56.1$$

The upper limit of acceptable CLI = 64. Therefore, this system passes the test.

APPENDIX B

Transmission Line Calculations

Velocity of propagation

$$\text{Velocity} = \frac{\text{distance}}{\text{time}}$$

Velocity of propagation of electrical energy along a transmission line of length D meters is given by

$$v_P = \frac{D}{\sqrt{LC}}$$

L = inductance of line henrys
C = capacitance of line farads
v_P = meters per second

Characteristic impedance of a transmission line

A schematic diagram of a section transmission line is shown below.

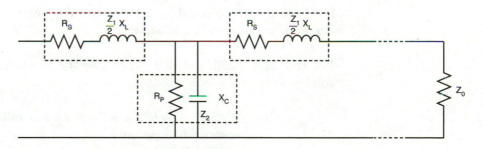

R_S is ac series resistance in ohms caused by wires.
X_L is a series inductive reactance caused by twisted wires.
R_P is insulation leakage resistance between conductors.
X_C is capacitive reactance caused by capacitance between conductors.

It can be shown that $Z_0 = \sqrt{Z_1 Z_2}$

Where Z, is the impedance of 2 (R_S in series with X_L)

Also, it can be shown that if $R_S = 0$ - $R_P = \infty$ for the lossless transmission line

$$Z_0 = \sqrt{\frac{L}{C}}$$

Therefore as series inductance is added by loading the pair, the characteristic impedance of the line is increased.

Given a transmission line shown in the diagram below.

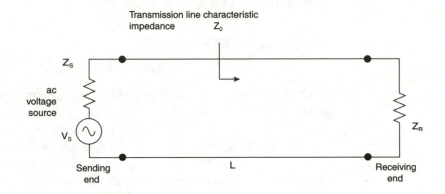

If Z_S and Z_R do not equal Z_0, voltage and current reflections will result along the transmission line length L.

The reflection coefficient K is defined as the magnitude of the reflected voltage wave to the incident voltage wave.

$$K = \frac{|V^-|}{|V^+|}$$

V− RMS voltage of the reflected wave, reflected from the mismatch impedance.

V+ RMS volts of the incident wave going toward the mismatch impedance.

Example: If V− = 0 V there is no reflected wave and K = 0.

K in terms of the impedance mismatch is represented as:

At the sending end $\qquad K_S = \dfrac{Z_S - Z_0}{Z_S + Z_0}$

At the receiving end $\qquad K_R = \dfrac{Z_R - Z_0}{Z_R + Z_0}$

Example: For a perfect impedance match $Z_O = Z_R = Z_S$

SO $K_S = 0$ K_R 0   No reflections

Since V⁺ is the incident voltage wave traveling down the line & V⁻ is the reflected wave traveling up the line caused by the mismatch of the receiving end impedance. V MAX occurs when they reinforce each other to cause a voltage maximum.

|ABSOLUTE| a) $|V_{MAX}| = |V^+| + |V^-|$ Cancel
  VALUE
   And   b) $|V_{MIN}| = |V^+| + |V^-|$ V⁺ & V⁻   each other.

The voltage standing-wave ratio is defined as:

$$VSWR = \frac{|V_{MAX}|}{|V_{MIN}|}$$

Substituting a.) & b.) in above

$$c.\ \ VSWR = \frac{|V^+| + |V^-|}{|V^+| - |V^-|}$$

Multiplying c.) by $\dfrac{|V^+|}{\dfrac{1}{|V^-|}}$

$$VSWR = \frac{1 + \dfrac{|V^-|}{|V^+|}}{1 - \dfrac{|V^-|}{|V^+|}} \qquad and \quad K = \frac{|V^-|}{|V^+|}$$

$$VSWR + \frac{1 + K}{1 + K} \qquad K\ can\ be\ K_R\ or\ K_S$$

Return loss is a measure of the incident wave power to the reflected wave power mathematically

$$Return\ loss \quad RL = 10 \log \frac{|P^+|}{|P^-|}$$

$$and\ |P^+| = \frac{|V^+|^2}{Z_o}, \ \ |P^-| = \frac{|V^-|^2}{Z_o}$$

Substituting

$$RL = 10 \log \frac{\dfrac{|V^+|^2}{Z_o}}{\dfrac{|V^-|^2}{Z_o}} = 10 \log \frac{|V^+|^2}{|V^-|^2}$$

$$RL = 20 \log \frac{|V^+|}{|V^-|} \quad \text{Since } K = \frac{|V^-|}{|V^+|}$$

$$RL = 20 \log \frac{1}{K}$$

Return loss is the difference expressed in dB between the incident and reflected waves along a transmission line.

APPENDIX C

SONET Networks and Testing Techniques

This appendix was made available through the courtesy of CERJAC, a subsidiary of Hewlett Packard, and is reprinted by permission. The material was chosen because it is extremely well done. The treatment of the subject is straightforward and easy to read, which makes this complicated topic become clear. SONET Network and Testing Technology was developed as a seminar and was presented at various technical conventions and expositions. Therefore there is a continuous progression from the present wired system to the Synchronous Optical Network (SONET). The parallel references to a truck transportation method of delivery is useful in illustrating SONET technology for transporting messages through the optical network.

This appendix provides an excellent overall view of typical network functions and implementations. Signals from various sources arrive at a SONET terminal through ATM, Frame Relay, and other types of networks. The digital data is then packetized according to SONET protocol to make up the payload sections of the synchronous frames. The compatibility of SONET with other types of networks is made clear. Therefore the adding and dropping of data from the SONET system is similar to the on-off ramps of a highway. This appendix provides an excellent, easy way to learn the elements of SONET networks and testing procedures.

Overview of SONET Technology

CERJAC Telecom Operation
43 Nagog Park
Acton, MA 01720

SONET/ATM
Networks and Testing Seminar

Abstract

The Synchronous Optical Network (SONET) is based on a worldwide standard for fiber-optic transmission, modified for North American asynchronous rates. This part of the seminar will cover the technology, concepts and philosophy of SONET. It will cover how SONET works in providing a migration path for existing asynchronous transmission technology. We will discuss the basic SONET technology, and the advantages it offers to both customers and service providers. Some of the issues that arise in deploying a SONET network will also be highlighted.

Author

CERJAC Telecom Operation
43 Nagog Park
Acton, MA 01720

Hewlett-Packard

Overview of SONET Technology

The objective of this seminar is to give a good understanding of SONET transmission technology and how it is being used in telecommunications in the 1990's.

Seminar Content

- Introduction to SONET concepts

- Basic SONET network devices

- Typical network architectures

- The SONET standards

- How SONET works

- Summary

This part of the seminar will cover the technology, concepts and philosophy of SONET. It will cover how SONET works in providing a migration path for existing asynchronous transmission technology. We will discuss the basic SONET technology, and the advantages it offers to both customers and service providers. Some of the issues that arise in deploying a SONET network will also be highlighted.

What is SONET?

Synchronous Optical NETwork

A transmission technology for fiber-optic transmission standardized worldwide which is replacing proprietary point-to-point transmission systems.

The Synchronous Optical Network (SONET) is based on a worldwide standard for fiber-optic transmission, modified for North American asynchronous rates. In the rest of the world, this technology is known as Synchronous Digital Hierarchy (SDH) which is designed to transport CEPT asynchronous transmission rates.

Why SONET?

- Transport system based on worldwide standard
  - –Compatible with existing asynchronous equipment
  - –Carry multiple tributaries of different types
  - –Internetworking between service providers

- Synchronous timing
  - –Direct access to tributaries
  - –DS3, DS1, DS0

- Includes OAM&P overhead
  - –Network management and maintenance support built-in

- Growth for future services
  - –ATM

The idea of SONET grew from the concerns of network operators at the lack of standards covering new Fiber-Optic Transmission Systems (FOTS). They saw that these proprietary designs, mainly for DS1 and DS3 transmission, made interworking and signal handover virtually impossible. The T1X1 committee of the Exchange Carriers Standards Association (ECSA) took up the challenge of creating standards.

The SONET standards we have today are the result of the cooperative effort of many interested groups from both private industry and other standards bodies. For example, in 1984 MCI proposed to the Inter-exchange Carrier Compatibility Forum (ICCF) the concept of standardizing the "mid span meet", the handover point of an optical transmission system. This proposal was passed on to ECSA for discussion and elaboration. In 1985 Bellcore proposed the "SONET" concept which was expanded on within the ECSA T1X1 committee. This was submitted to ANSI (American National Standards Institute) for endorsement and released in June 1988 as the ANSI T1.105 standard in use today.

SONET network equipment will also give savings in cost and size. Using a concept called "Direct Synchronous Multiplexing", there is no need to de-multiplex tributary channels before switching, as must be done with the existing DS1/DS3 asynchronous network. By having the optical network operate synchronously, this new concept in multiplexing makes possible circuit re-routing in "real-time".

A further consideration to compatibility is at the network management level. SONET standards include many performance monitoring and reporting systems. By defining the message sets for communicating with Operation Support Systems (OSS), network information may be passed from one network operator to another where desired, and different vendors' NEs will provide the same information to the OSS.

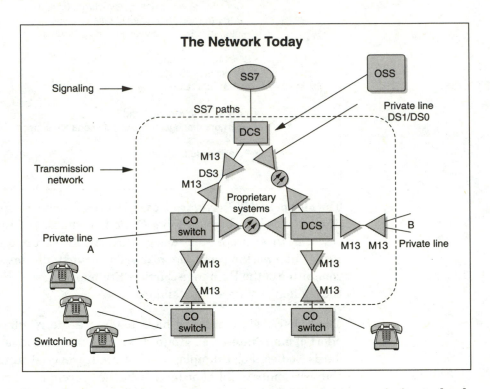

Typically, in existing networks, simple point-to-point transmission technology is used to link the network switches or customer locations. A DS0 signal (64 kbit/s) from a phone call, for example, may be multiplexed up to DS1 using a D4 or D5 channel bank and then to DS3 using an M13 multiplexer. However, to switch the DS0 signal, the full DS3 signal must be demultiplexed. This requires a full set of muxes at each end of a DS3 transmission link. This is also true for higher rate proprietary fiber-optic transmission systems (FOTS). This back-to-back arrangement is very expensive, when in practice only some of the lower-order signals need switching.

Some circuits may use manual "patch panels" rather than Digital Cross-connect Systems (DCS) to route a circuit through the network. Re-provisioning in the event of a customer no longer needing a facility is very time consuming and expensive if re-patching is needed or equipment has to be re-located or retrieved. Even with existing DCS systems, re-routing circuits can take from minutes to hours depending on the control methods.

With a SONET network, all bandwidth allocation and transmission routing can be controlled remotely, making it simple to re-route or re-provision circuits. This can be performed within the SONET overhead, making it easier and less expensive to reconfigure the network "on-the-fly".

The primary issues with today's network are:

- higher rate line systems are proprietary,
- the network architecture is inflexible and expensive for newer services, and
- there is limited built-in network management and maintenance support capabilities

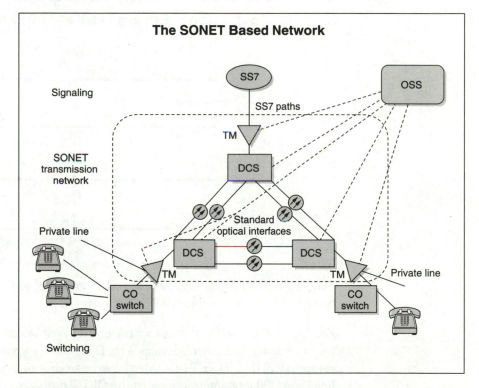

The SONET Based Network

The SONET network has exactly the same function as the existing asynchronous DS3 based network i.e., it will transport customer data from one location to another. However, the SONET standards define transmission rates to 2.4 Gbit/s. This is a massive increase from the existing DS3 rate of 45 Mbit/s.

New bandwidth intensive services like High Definition TV, ultra high-speed computer links, Local Area Network (LAN) bridge links and Broadband

ISDN will all be possible. In theory, the SONET network will be viewable as a transmission "cloud". A customer need only be connected to the nearest "Access Device" at the locations between which he requires service.

The OSS provisioning system will then define circuit paths between the locations making use of spare capacity on the SONET transmission systems. With a software-configurable network, in the event of network failure, the routing may be changed within milliseconds.

With this ability to re-route, SONET will take care of equipment failures with negligible effect on customer service.

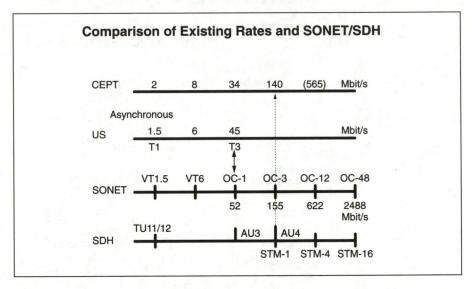

Comparison of Existing Rates and SONET/SDH

SDH stands for "Synchronous Digital Hierarchy" and is the ITU (CEPT) international standard for SONET transmission.

For SONET to be compatible with existing technology it must be able to carry existing customer services e.g., DS1, DS3. As you can see from the comparison, the SONET line rates have been chosen to match at the lowest level the DS3 customer service. In the CEPT multiplex hierarchy, used outside the USA, a rate of 155 Mbit/s was chosen to match the 140 Mbit/s standard for high capacity bearer transmission.

At present, SONET and SDH standards define signal structures up to 2.4 Gbit/s (OC-48). Higher rates at 4.8 Gbit/s (OC-96) and 9.6 Gbit/s (OC-192) are currently being investigated.

NOTE: The SONET "Optical Carrier" (e.g., OC-3) definitions translate to "Synchronous Transport Module" types (e.g., STM-1) in the SDH standards.

The transmission capability afforded by SONET is massive compared to previous fiber-optic transmission speeds. The SONET standards define FOTS as Optical Carrier (OC) types from OC-1 for existing DS3 service to OC-48 which is a 2.4 Gbit/s FOTS.

SONET Signal Hierarchy and Line Rates

Synchronous transport signal	Line rate Mbit/s	Optical carrier
STS-1	51.84*	OC-1
STS-3	155.52*	OC-3
STS-12	622.08*	OC-12
STS-24	1244.16*	OC-24
STS-48	2488.32*	OC-48

*Denotes the popular SONET physical layer interfaces

The lowest level SONET signal is called the Synchronous Transport Signal level 1 (STS-1) which has a signal rate of 51.84 Mbit/s. The optical equivalent of the STS-1 is the Optical Carrier level 1 signal (OC-1) which is obtained by a direct electrical to optical conversion of the STS-1 signal.

Higher level signals, obtained by byte-interleaved multiplexing lower level signals, are denoted by STS-N and OC-N, where N is an integer. The line rate of the higher level OC-N signal is N times 51.84 Mbit/s, the line rate at the lowest level.

The SONET standard allows only certain values of N. These values of N are: 1, 3, 9, 12, 18, 24, 36, and 48. Values of N greater than 48 may be allowed in future revisions. A maximum value of 255 is allowed under the current standard.

At present, the important SONET line signals and rates are OC-1 at 51.84 Mbit/s, OC-3 at 155.52 Mbit/s, OC-12 at 622.08 Mbit/s, and OC-48 at 2.488 Gbit/s.

Who is using SONET?

- Telephone companies
  - RBOCs, independents

- Long haul carriers
  - AT&T, MCI, Sprint, Wiltel

- Competitive access providérs
  - MFS, Teleport

- Private network operators
  - Power companies, Boeing Computer Services

- High-bandwidth end-users
  - VISTA net

A wide range of telecom service providers will benefit from installing SONET technology. From existing telephone companies and private network operators to the operators of high speed city wide MANs, all will eventually move to SONET. The benefits of using SONET are many.

SONET technology is becoming the primary transmission technology employed for telecommunications services; most telecommunications service providers are now installing SONET equipment. Existing phone companies, private network operators, competitive access providers, and operators of high capacity WAN/MANs have embraced SONET. New services, particularly ATM require SONET and are increasing demand further.

All major equipment vendors make SONET transmission and multiplex technology.

Why do they want it?

- Common standard—multi-vendor interoperability

- Better management on OSS

- Fast provisioning

- Better network survivability

- Simpler handover

- Support for future services

- COST

Interoperability: With a common standard, compatible FOTS equipment will be available from many vendors. In a highly competitive market prices will be very attractive. Also, the service provider can easily network with other service providers.

Better network management: With better network management, operators will be able to more efficiently use the network and provide better service. The concept of OSS (Operations Support Systems) is under study by Bellcore. Some OSS standards defining management system interfaces already exist.

Faster provisioning: If new circuits can be software defined to use existing spare bandwidth then provisioning will be much faster. The only new connection needed will be from the customer's premises to the nearest network access node.

Better network survivability: With "real-time" re-routing possible the OSS will be able to take care of failure by simply reprogramming circuit paths. The built-in SONET protection and reporting systems will automatically take care of simple transmission failures.

Simpler handover: If all networks use equipment to the same standard the handover of circuits at the "mid-span meet" should be trouble free.

Support of future services: Looking to the future, the SONET design will cater for new services like High Definition TV, CAD/CAM systems, Wide Area Network backbone networks, Broadband ISDN and new bandwidth-on-demand services. As the SONET operator will have total control of bandwidth allocation, any new service will be simple to provision.

COST: SONET provides a system that is cheaper to install, maintain and provide a given level of service. It will allow the network providers to increase their services while maintaining or lowering their costs.

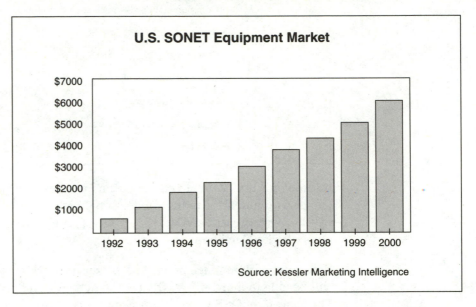

U.S. SONET Equipment Market

Source: Kessler Marketing Intelligence

Network operators are now installing SONET equipment rapidly with many field trials and the appearance of more equipment manufacturers. Most transmission equipment purchased for US networks is now SONET, with new services (e.g. ATM) continuing to drive growth of SONET.

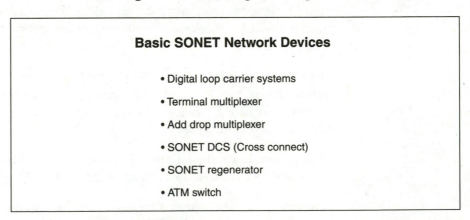

Basic SONET Network Devices

- Digital loop carrier systems
- Terminal multiplexer
- Add drop multiplexer
- SONET DCS (Cross connect)
- SONET regenerator
- ATM switch

Having introduced you to the concept of a SONET network, let's now take a look at the network "building blocks" and how they are configured.

These network elements are now all defined in Bellcore or ANSI standards and provide multiplexing or switching functions. These elements (with the exception of add-drop multiplexers) have the same functions as asynchronous transmission equipment.

Loop Carrier Systems (DLC)–Digital Loop Carriers are specialized SONET back-to-back mux systems providing circuit concentration in the local loop market. The elements used are similar to the Terminal Mux but transmission speed is normally limited to 155 Mbit/s (OC-3).

Terminal Mux–simple multiplexing of SONET and standard DS1/DS3 channels onto a single SONET bearer. Usually described as part of a FOTS/FTS if operating at 622 Mbit/s (OC-12) or above.

Add Drop Mux–a terminal multiplexer with the ability to operate in through mode (ADM) and add channels to, or drop channels from, the through signal. This may be used to add, drop or cross-connect tributary channels. They may operate at any SONET rate.

SONET DCS–full cross connect capability at SONET rates using "Direct Synchronous Switching".

Regenerator–for SONET transmission over about 35 miles, regenerators are required with spacing dependent on the transmission technology. These are not just simple signal reconstituters but have alarm reporting and error reporting capability.

ATM Switch–a cell-based switch that may interface to other switches (Network-to-Network Interface or NNI) or user devices (User-to-Network Interface or UNI).

Every network element has alarm reporting and performance monitoring, allowing a fault to be isolated quickly to the individual transmission section with the problem.

Typical Network Architectures

- Point-to-point

- Multi-point/Linear drop

- Rings

- Bi-directional line switched rings

- Path switched ring

- Interconnected SONET islands

SONET technology is being deployed in new installations and to replace or upgrade existing systems when they reach capacity. Using the basic SONET network devices, a variety of network topologies are being used to provide services to customers. These topologies include:

- point-to-point systems,
- linear multi-point systems,
- rings, and
- interconnected SONET networks

Point-to-point

At the simplest level new point-to-point systems will use SONET Terminal Muxes with the ability to expand to more complex SONET constructions later.

This arrangement is similar to that used with M13 muxes in the present asynchronous network. There are two versions of point-to-point systems based on the SONET network elements described earlier:

1. The Digital Loop Carrier system (DLC). These will be used in the local loop to convey concentrated customer circuits to the CO switch. A remote data terminal (RDT) has essentially the same function but the CO mux may have greater management and fault reporting capability. The RDT must also deal with customer interfaces by signaling.

2. Point-to-Point Terminal Multiplexers. These will provide transmission within the SONET network in a similar manner to the M13 muxes today but at higher data rates.

Multi-point/linear drop

An extension of the point-to-point is the multi-point where additional TMs are placed in the point-to-point path to add traffic at new locations. This is where SONET allows flexibility beyond that available with the existing DS3 systems. As SONET multiplexing is simpler this is likely to be a less expensive solution than available with DS3 based systems.

Rings

The ring is a most important new topology possible with SONET. SONET capability goes beyond that of existing systems with the Add-Drop Mux. With the ability to add and drop channels at any ADM, under centralized management control, a very flexible network architecture is possible. Combine this with the reliability of a bi-directional ring topology and a very robust and readily managed network is possible.

Two types of rings have been used in SONET, the path switched ring, and the bi-directional line switched ring.

The path switched ring provides protection switching at the path level. If a failure occurs, the lower level path signal is switched to the protection fiber. Thus live traffic could be carried on both fibers.

In a line switched ring, all traffic at the SONET line level is switched to a protection fiber during a fault. All traffic is normally carried on one of the fibers in the ring.

Interconnected Networks

As greater amounts of SONET equipment is deployed, the network will become highly interconnected. Rings, linear systems and meshes will all be part of the network in which "islands" of SONET equipment are interconnected. These interconnections will lead to many internetworking issues we will discuss.

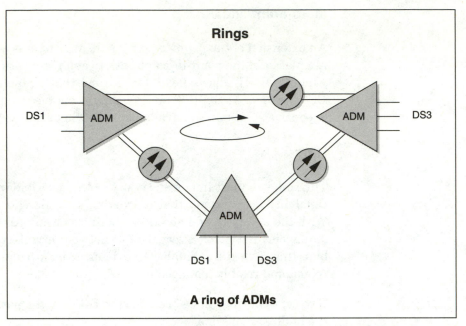

A ring of ADMs

Rings are becoming a key architecture of SONET systems. Should one fiber route be broken, the ADMs can automatically reroute traffic in the opposite direction around the ring. Rings also make it easier to manage services and bandwidth due to the many locations and possible circuit routes to them. Rings are already in service in major metropolitan areas with many new rings being planned and installed.

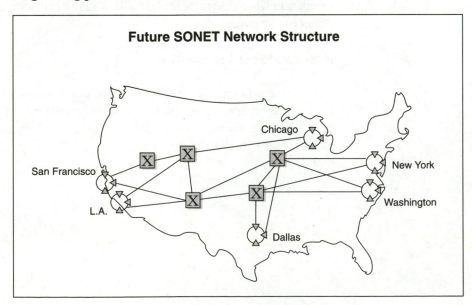

With simplified handover from LATAs to Longhaul Carriers future SONET networks throughout the US may look like this. At each "mid span meet" the interworking of different vendors equipment should be assured if the equipment complies to the standards. However, there will likely be mis-interpretations of the standards which will require test equipment to resolve.

Although SONET equipment may be compatible for transmission, it will only be when the management messages have been fully implemented that true interworking will be possible.

The SONET Standards

Bellcore (800)	TR-253	Generic requirements
521 CORE	TR-233	Wideband and broadband DCS
	TR-303 Loop carriers	
	TR-496 ADM	
	TR-499 Transport system	generic requirements
	TR-782 SONET digital	trunk interface
	TR-917 SONET	regenerator equipment
ANSI	T1.105.xx	Rates, formats, jitter, etc.
	T1.106 Optical interface	
	T1.119 OAM&P	communications
	T1.204 Operations,	administration and provision
	T1.231 In-service	performance monitoring

The driving force behind SONET is the standards which define the network equipment functionality and performance. The original Bellcore standard was released in 1985 as TR-TSY-00253. This has been re-issued many times as new requirements are added. GR-TSY-00253 is now in preparation. Other Bellcore standards appeared in November 1989, e.g.. TR-TSY-000496 which defined the operation of the ADM and TR-TSY-000499 (Transport Systems General Requirements).

AT&T and others followed Bellcore in releasing design standards for their own versions of a synchronous network. The SONET submissions have now been amalgamated by ANSI committees into a number of standards, e.g.:

ANSI T1.105 —Digital Hierarchy —Optical Interface Rates and Formats Specification.
ANSI T1.106 —Digital Hierarchy —Optical Interface Specifications.

The T1.105.xx document is now in the process of being updated and revised. ANSI is the coordinating/endorsement body for many approved committees like the Exchange Carriers Standards Association (ECSA) T1X1 committee. The specifications defined by the committees, if approved, are released as ANSI standards.

ANSI (American National Standards Institute) has planned and released the SONET standards in defined Phases. The Phase 1 release covered all hardware operation and interfacing. Phases 2 and 3 covered the management protocols and messages. The reasoning here was that as the protocols are software based it would be simply a matter of software upgrades to bring the already installed SONET equipment up to Phase 3 capability. Manufacturers have the task of upgrading the software to the new standards.

The US SONET standards required some modifications to permit interworking with the networks of other countries. These modifications were incorporated into CCITT standards covering SDH (Synchronous Digital Hierarchy) which is the international equivalent of SONET. The primary standards are CCITT G.707-709.

These standards define the operation of network elements down to the line signal formats, interfaces and security features. It is the in-depth operation of the SONET transmission we will examine next.

How SONET Works

- Overview
  - transmission rates
  - terminology

- The SONET frame structure

- Concept of synchronous frames

In this section we will take a look at the SONET signal from the transmission rates involved to the concepts and structure of the transmitted data. We will describe how an existing asynchronous signal is transported over a SONET network.

SONET Terms

OC-n	Optical carrier n (n = 1 thru 48)
STS-n	Synchronous transport signal n (n = 1 thru 48)
EC-n	Electrical carrier n (n = 1, 3)
SPE	Synchronous payload envelope —carrier user data
VT	Virtual tributary —subdivision of SPE for lower rates

n = Multiples of basic 51.84 Mbit/s transport signal

SONET, like any other new technology, has many new descriptions, abbreviations and terms. This slide lists the main SONET abbreviations which relate to the SONET transmission concepts:

OC-n —Optical Carrier type "n", where n as we have seen is from 1 (51.84 Mbit/s) to 48 (2.4 Gbit/s). Most popular line signal rates are OC-3 at 155.52 Mbit/s, OC-12 at 622.08 Mbit/s, and OC-48 at 2.488 Gbit/s.

EC-n —Electrical Carrier-n, where n is 1 or 3. This is the electrical interface defined for the 51.84 Mbit/s and 155.52 Mbit/s signal.

STS-n —Synchronous Transport Signal "n". This is the SONET "transport" data structure appearing on the OC-n" FTS. The STS consists of "frames" into which data is filled.

SPE —Synchronous Payload Envelope. A defined area within the STS-N which carries the data for customer services.

VT —Virtual Tributary signal. An SPE can be defined to carry different services. Depending on the data rate of the services the data area within the SPE will be allocated —"Mapped" —to suit. Within the SPE a service will have a defined area, a Virtual Tributary frame.

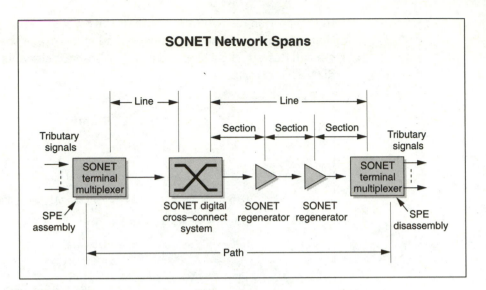

For network management and maintenance purposes, the SONET network may be described in terms of three different network spans:

1. The PATH span which allows network performance to be maintained from a customer service end-to-end perspective;

2. The LINE span which allows network performance to be maintained between transport nodes and provides the majority of network management reporting; and

3. The SECTION span which allows network performance to be maintained between line regenerators or between a line regenerator and a SONET Network Element allowing fault localization.

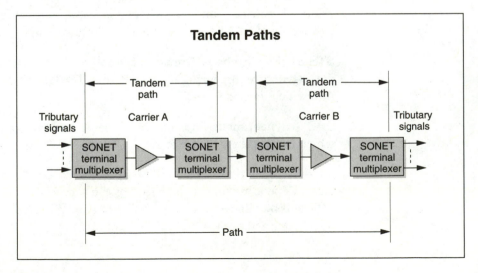

The transport of a customer's signal often involves multiple service providers. As SONET evolved, a need arose to provide a tandem monitor function for use by service providers. Spare overhead capacity (Z5 byte) in the SONET path overhead has now been defined for use in Tandem Connection Maintenance (T1.105.05-1994).

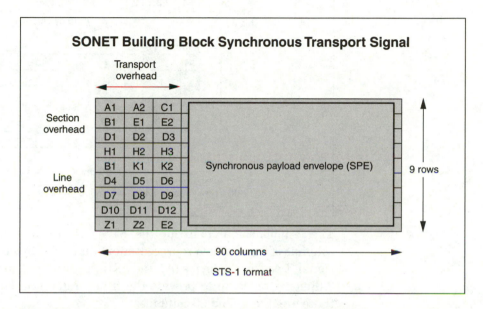

The basic building block of SONET transport is the Synchronous Transport Signal (STS). For clarity, the bits that are transmitted sequentially in the SONET frame are represented in a 2 dimensional matrix of 810 (9 × 90) bytes. 27 of the bytes are the transport overhead, the rest are the Synchronous Payload Envelope (SPE).

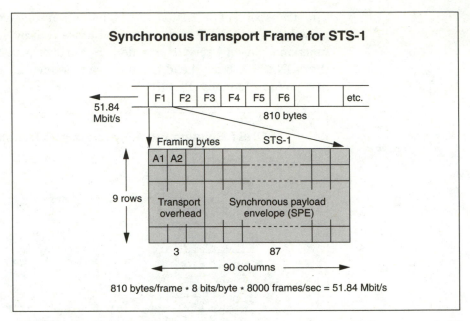

The 2-dimensional map for the STS frame comprises N rows and M columns of boxes which represent individual 8-bit bytes of the synchronous signal. Two framing bytes (A1 and A2) appear in the top left hand box of the 2-dimensional map to provide the frame reference. For an STS-1 frame, there are 9 rows and 90 columns.

The concept of transporting tributary signals intact across a synchronous network has resulted in the term "Synchronous Transport Frame" being applied to such synchronous signal structures. More importantly, however, signal capacity is set aside within a synchronous transport frame to support network transportation capabilities. A synchronous transport frame, therefore, comprises two distinct and readily accessible parts within the frame structure —a synchronous payload envelope part and a transport overhead part.

Synchronous Payload Envelope (SPE): Individual tributary signals (such as a DS3 signal for example) are arranged within the "Synchronous Payload Envelope" which is designed to traverse the network from end-to-end. This signal is assembled and disassembled only once even though it may be transferred from one transport system to another many times on its route through the network.

Transport Overhead (TOH): Some signal capacity is allocated in each transport frame for "Transport Overhead" which provides the facilities (such as alarm monitoring, bit-error monitoring and data communications channels) required to support and maintain the transportation of an SPE between nodes in a synchronous network. Transport overhead pertains only to an individual transport system and is not transferred with the SPE between transport systems.

Scrambling: To ensure that the clock can always be recovered from the received data all but the framing and C1 bytes within the SPE are scrambled.

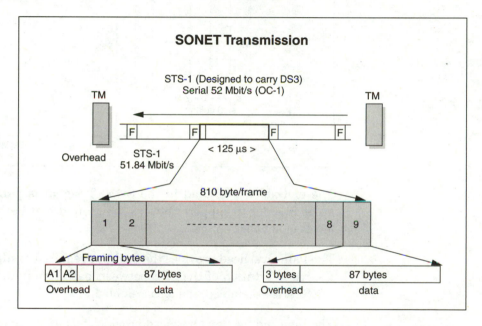

In each frame, the signal bits are transmitted in a sequence starting with those in the top left hand corner byte, followed by those in the 2nd byte in row 1, and so on, until the bits in the 90th (last) byte in row 1 are transmitted. Then, the bits in the 1st byte of row 2 are transmitted, followed by the bits in the 2nd byte of row 2, and so on, until the bits in the 90th byte of the 2nd row are transmitted. The sequence continues with the bytes in the 3rd row followed by the bytes of the 4th row, and so on, until the 90th byte of the 9th row is transmitted. Then the whole sequence repeats.

The synchronous signal comprises a set of 8-bit bytes which are organized into a frame structure. Within this frame structure, the identity of each byte is known and preserved with respect to framing or marker bytes.

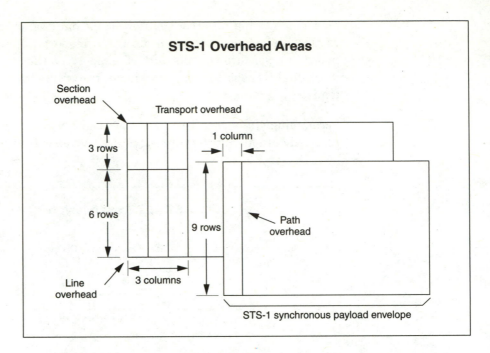

STS-1 Overhead Areas

The embedded overhead in the SONET signal is partitioned into three areas in order to support network maintenance at Path, Line and Section levels of network spans.

The Path Overhead provides the facilities required to support and maintain the transportation of the SPE between path terminating locations where the SPE is assembled and disassembled.

The Line and Section Overhead provides facilities to support and maintain the transportation of the SPE between adjacent nodes in the SONET network. These facilities are included within the Transport Overhead part of the transport frame. Thus, the STS-1 Transport Overhead, comprising the first 3 columns of the STS-1 frame, is split between Section Overhead and Line Overhead.

The Section Overhead occupies the top 3 rows of the Transport Overhead for a total of 9 bytes in each STS-1 frame. The Line Overhead occupies the bottom 6 rows of the Transport Overhead for a total of 18 bytes in each STS-1 frame.

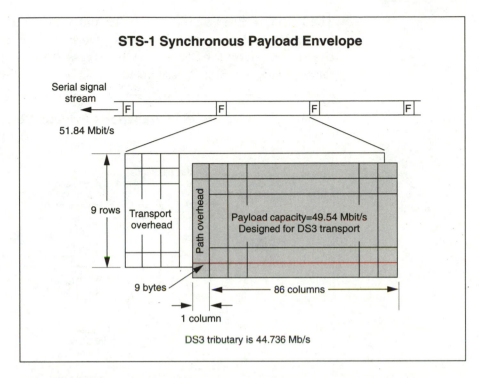

The Synchronous Payload Envelope comprises two parts: a Payload Capacity part and a Path Overhead part.

Payload capacity:

The Payload Capacity area of each SPE is intended to support the transportation of specific tributary signals. The STS-1 payload capacity comprises 774 bytes, structured as 86 columns of 9 bytes. These bytes provide a 49.54 Mbit/s transport capacity with a frame repetition rate of 8000 Hz. This payload capacity has been designed specifically to accommodate the transportation of a DS3 tributary signal (at 44 Mbit/s).

Path overhead:

An area of each SPE is also allocated for Path Overhead. This signal capacity provides the facilities (such as alarm monitoring and performance monitoring) required to support and maintain the transportation of the SPE between end locations (known as Path Termination's) where the SPE is either assembled or disassembled. It also supports the Tandem Connection Maintenance functions. Signal capacity for the STS-1 SPE Path Overhead is allocated in the first column of the STS-1 SPE —a total of 9 bytes per frame.

The Path Overhead comprises 9 bytes and occupies the first column of the
Synchronous Payload Envelope. Path Overhead is created and included in
the SPE as part of the SPE assembly process. It remains as part of the SPE
for as long as the SPE is assembled.

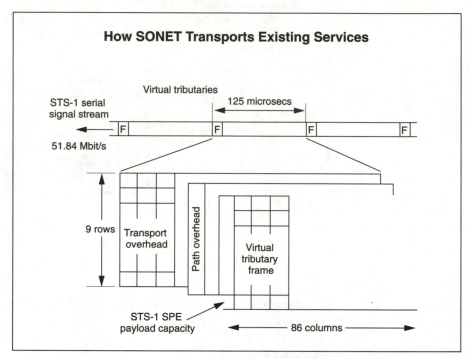

The SONET STS-1 SPE with a channel capacity of 50.11 Mbit/s has been
designed specifically to provide transport for a DS3 tributary signal. Transport for a tributary signal with a signal rate lower than that of a DS3, such
as a DS1 for example, is provided by a Virtual Tributary (VT) frame structure.

VTs are specifically intended to support the transport and switching of payload capacity which is less than that provided by the STS-1 SPE. By design,
the VT frame structure fits neatly into the STS-1 SPE in order to simplify
VT multiplexing capabilities. A fixed number of whole VTs may be assembled within the STS-1 SPE.

The concept of a tributary signal (such as a DS3 signal) being assembled
into a Synchronous Payload Envelope, to be transported end-to-end across
a synchronous network, is fundamental to the SONET standard. This
process of assembling the tributary signal into an SPE is referred to as
"payload mapping".

To provide uniformity across all SONET transport capabilities, the payload capacity provided for each individual tributary signal is always slightly greater than that required by the tributary signal. Thus, the essence of the mapping process is to synchronize the tributary signal with the payload capacity provided for transport. This is achieved by adding extra stuffing bits to the signal stream as part of the mapping process.

Thus, for example, a DS3 tributary signal at a nominal rate of 44 Mbit/s needs to be synchronized with a payload capacity of 49.54 Mbit/s provided by the STS-1 SPE. Addition of the Path Overhead completes the assembly of the STS-1 SPE and increases the bit rate of the composite signal to 50.11 Mbit/s.

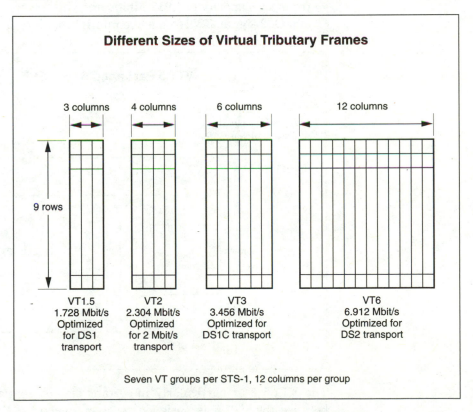

Different Sizes of Virtual Tributary Frames

3 columns — VT1.5 — 1.728 Mbit/s Optimized for DS1 transport

4 columns — VT2 — 2.304 Mbit/s Optimized for 2 Mbit/s transport

6 columns — VT3 — 3.456 Mbit/s Optimized for DS1C transport

12 columns — VT6 — 6.912 Mbit/s Optimized for DS2 transport

9 rows

Seven VT groups per STS-1, 12 columns per group

A range of different VT sizes is provided by SONET:

(i)   VT1.5: Each VT1.5 frame consists of 27 bytes, structured as 3 columns of 9 bytes. At a frame rate of 8000 Hz, these bytes provide a transport capacity of 1.728 Mbit/s and will accommodate the mapping of a 1.544 Mbit/s DS1 signal. 28 VT1.5s may be multiplexed into the STS-1 SPE.

(ii)   VT2: Each VT2 frame consists of 36 bytes, structured as 4 columns of 9 bytes. At a frame rate of 8000 Hz, these bytes provide a transport capacity of 2.304 Mbit/s and will accommodate the mapping of a CEPT 2.048 Mbit/s signal. 21 VT2s may be multiplexed into the STS-1 SPE.

(iii)   VT3: Each VT3 frame consists of 54 bytes, structured as 6 columns of 9 bytes. At a frame rate of 8000 Hz, these bytes provide a transport capacity of 3.456 Mbit/s and will accommodate the mapping of a DS1C signal. 14 VT3s may be multiplexed into the STS-1 SPE.

(iv)   VT6: Each VT6 frame consists of 108 bytes, structured as 12 columns of 9 bytes. At a frame rate of 8000 Hz, these bytes provide a transport capacity of 6.912 Mbit/s and will accommodate the mapping of a DS2 signal. 7 VT6s may be multiplexed into the STS-1 SPE.

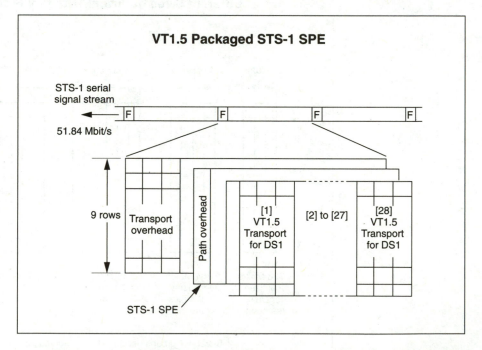

The VT1.5 is a particularly important size of virtual tributary. This is because the VT1.5 is designed to accommodate a DS1 tributary signal which has the highest density of all the tributary signals that appear in the existing network. Twenty-eight VT1.5s can be packaged into the STS-1 SPE for transportation.

The 3 column by 9 row structure of the VT1.5 fits neatly into the same 9 row structure of the STS-1 SPE. Thus, twenty-eight VT1.5s may be packed into the 86 columns of the STS-1 SPE payload capacity. This leaves 2 columns in the STS-1 SPE payload capacity spare. These spare columns are filled with fixed stuff bytes which allows the STS-1 SPE signal structure to be maintained.

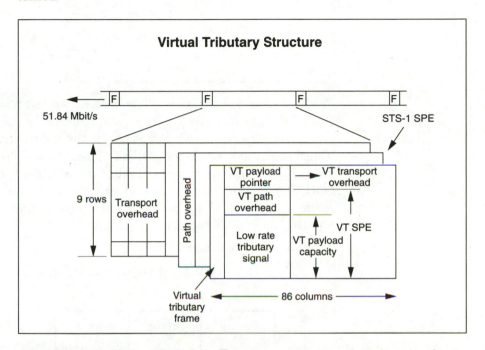

Essentially, the Virtual Tributary Frame represents a mini transport frame structure. It has the attributes of a SONET Transport Frame but is carried within the standard SONET STS-1 Frame structure.

Thus a low rate tributary signal may be mapped into the VT Payload Capacity. VT Path Overhead is added to this payload capacity to complete the VT Synchronous Payload Envelope (VT SPE). The VT SPE is linked to the VT frame by means of a VT Payload Pointer which is the only component of VT Transport Overhead. The VT frame is then multiplexed into a fixed location within the STS-1 SPE.

The VT frame structure is illustrated here as residing in one STS-1 SPE. In fact, however, this frame structure is distributed over four consecutive STS-1 SPE frames. It is, therefore, more accurate to refer to the structure of the VT as a VT multiframe or superframe. The phase of the multiframe is indicated by the H4 byte in the Path Overhead.

Network Functions of the Overhead Capabilities

- Section overhead

- Line overhead

- Path overhead

- VT overhead

The final section of our study of how SONET works considers the embedded overhead capabilities within the signal. The high level of network management possible with SONET depends on the information provided by the "Overhead" areas within the SONET STS-n frames. Typically within the 3 defined overhead areas called Path, Line and Section Overhead (POH, LOH, SOH) there are transmission error detection and reporting features, communication channels, pointers and frame content codes. There is also VT Overhead for VT transport.

Section Overhead Bytes

Section overhead	Framing A1	Framing A2	Framing A1	Path trace J1
	BIP-8 B1	Orderwire E1	User F1	BIP-8 B3
	Data com D1	Data com D2	Data com D3	Signal label C2
Line overhead	Pointer H1	Pointer H2	Pointer H3	Path status G1
	BIP-8 B2	APS K1	APS K2	User channel F2
	Data com D4	Data com D5	Data com D6	Multiframe H4
	Data com D7	Data com D8	Data com D9	Growth Z3
	Data com D10	Data com D11	Data com D12	Growth Z4
	Growth Z1	Growth Z2	Orderwire E2	Growth Z5

Path Overhead [carried in SPE]

The Section overhead consists of
    framing,
    ID,
    parity,
    orderwire,
    user, and
    data communications channel

Two bytes (A1 and A2) are assigned to the framing pattern in an STS-1 frame for frame alignment. N times 2 bytes, which are byte-interleaved, are assigned to the framing pattern of an STS-N frame. Each STS-1 within an STS-N is identified separately by a binary number corresponding to its order of appearance in the byte-interleaved STS-N frame (C1 byte). An 8-bit wide bit-interleaved parity check (B1 byte) is calculated over all bits of the STS-N frame. The computed value is placed in the Section Overhead of the following STS-N frame.

A 192 kbit/s Data Communications Channel (DCC) is provided (D1–D3 bytes). This DCC is intended to allow message-based network management and maintenance information to be exchanged between Section terminating equipment, such as regenerators and remote terminals.

An orderwire (E1 byte) is provided for voice communications between Section terminating equipment. It is intended that this channel shall be used as a local orderwire reserved for voice communication between regenerators, hubs, and remote terminal locations.

Signal capacity is provided for an additional User Channel for network operator communications (F1 byte). This channel is intended for use in proprietary data communications applications and will be terminated at each Section terminating equipment and passed on from one Section terminating equipment to another.

Section Overhead Functions

- A1, A2 — Frame alignment pattern

- C1 — STS-1 identification

- B1 — Parity check

- D1, D2, D3 — Data communications channel—192 kbit/s

- E1 — Voice communications [Orderwire]

- F1 — User channel

The 9 bytes of the STS-1 section overhead are made up as follows.

A1, A2: Two bytes (A1 and A2) provide a frame alignment pattern (11110110 00101000 binary, F6 28 hex). These bytes are provided in all STS-1s within an STS-N.

C1: The C1 byte is set to a binary number corresponding to its order of appearance in the byte interleaved STS-N frame and can be used in the framing and de-interleaving process to determine the position of other signals. This byte is provided in all STS-1s within an STS-N with the first STS-1 being given the number 1 (00000001).

B1: The B1 byte provides "section" error monitoring by means of a bit-interleaved parity 8 code using even parity (BIP-8). In an STS-N, the section BIP-8 is calculated over all bytes of the previous STS-N frame after scrambling and the computed value is placed in the B1 byte of STS-1 number 1 before scrambling.

E1: The E1 byte provides a local orderwire channel for voice communications between regenerators, hubs and remote terminal locations and is only defined for STS-1 number 1 of an STS-N signal.

F1: The F1 byte is allocated for user's purposes and is terminated at all section level equipment. This byte is defined only for STS-1 number 1 of an TS-N frame. It might be used to download firmware to regenerator.

D1-D3: The three bytes D1, D2 and D3 provide a data communications channel for message-based administration, monitor, alarm, maintenance and other communications needs at 192 kbit/s between section termination equipment. These bytes are defined only for STS-1 of an STS-N frame.

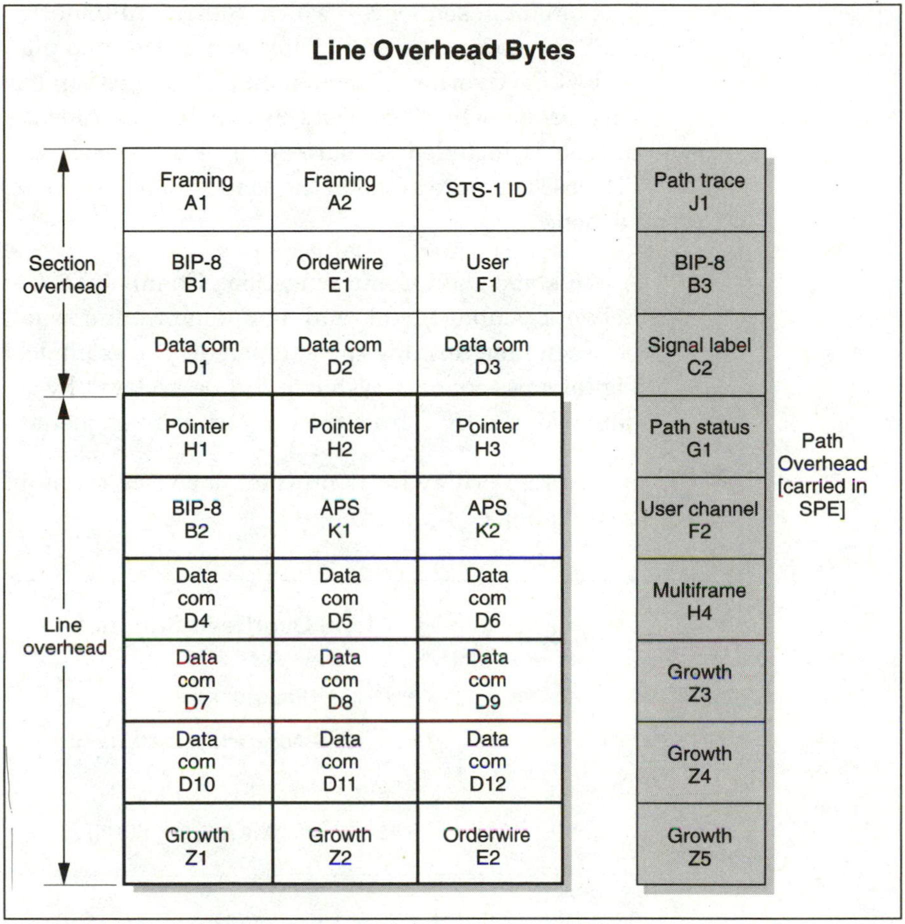

Line Overhead Bytes

Framing A1	Framing A2	STS-1 ID		Path trace J1
BIP-8 B1	Orderwire E1	User F1		BIP-8 B3
Data com D1	Data com D2	Data com D3		Signal label C2
Pointer H1	Pointer H2	Pointer H3		Path status G1
BIP-8 B2	APS K1	APS K2		User channel F2
Data com D4	Data com D5	Data com D6		Multiframe H4
Data com D7	Data com D8	Data com D9		Growth Z3
Data com D10	Data com D11	Data com D12		Growth Z4
Growth Z1	Growth Z2	Orderwire E2		Growth Z5

Section overhead (rows 1–3), Line overhead (rows 4–9)

Path Overhead [carried in SPE]

The Line Overhead consists of:

- payload pointer
- parity
- APS
- orderwire
- DCC
- growth (unassigned)

The Payload Pointer bytes which provide the linkage between the Transport Overhead and Synchronous Payload Envelope are processed as part of the Line Overhead functionality. They are passed unchanged by section terminating equipment. Separate Payload Pointers are provided for each STS-1 in an STS-N frame. Thus, the Line Overhead provides the necessary support for the networking capabilities of SONET.

The protocol sequences which control Automatic Protection Switching (APS) to alternate stand-by line equipment and plant are signaled as part of the Line Overhead functionality. This functionality provides further support for the SONET networking capabilities. Alarm and parity check information is included as part of the Line Overhead. (These functions are discussed in more detail in the context of in-service testing at the end of this section.)

A 576 kbit/s Data Communications Channel (DCC) allows message-based network management and maintenance information to be exchanged between Line terminating equipment. For example, the routing tables of a digital cross-connect system could be updated by sending the appropriate data over the DCC from the network management computer.

An express orderwire is provided for voice communications between line terminating equipment.

Line Overhead Functions

- H1-H3 — Payload pointers

- K1, K2 — Automatic protection switching

- B2 — Parity check

- K2 — Alarm (line AIS, line FERF)

The 18 bytes of the STS-1 line overhead are made up as follows.

H1-H3: The three bytes H1, H2 and H3 facilitate the operation of the STS-1 payload pointer and are provided for all STS-1s in an STS-N.

B2: The B2 byte provides a BIP-8 "line" error monitoring function. The line BIP-8 is calculated over all bits of the line overhead and payload envelope capacity of the previous STS-1 frame before scrambling and the computed value is placed in the B2 byte before scrambling. This byte is provided for each STS-1 in an STS-N.

K1, K2: The two bytes K1 and K2 provide Automatic Protection Switching (APS) signaling between line terminating equipment and are defined only for STS-1 number 1 in an STS-N. The K2 byte is also used to transmit performance information.

Line Overhead Functions

• D4-D12 — Data communications channel—576 kbit/s

• E2 — Voice communications (Orderwire)

• Z2 — Line FEBE and growth

• Z1 — Synchronization messages

D4-D12: The nine bytes D4 to D12 provide a data communications channel at 576 kbit/s for message-based administration, monitor, maintenance, alarm, and other communications needs between line termination equipment. These bytes are defined only for STS-1 number 1 of an STS-N.

E2: The E2 byte provides an express orderwire channel for voice communications between line terminating equipment and is only defined for STS-1 number 1 of an STS-N signal.

Z2: Part of the Z2 byte is used for line layer Far End Block Error (FEBE), which provides a count of the far end line B2 errors. The remaining bits in the Z2 byte are reserved for future use.

Z1: The Z1 byte has been assigned as a means of passing synchronization messages by the standards committees. The definitions are still in discussion.

Protection Switching

• APS messaging in K1, K2 bytes of first STS-1

• Linear APS

• Ring APS
  –Path switched
  –Line switched
  –Bidirectional line switched—T1.105.01

Protection switching is performed using the messaging capability built into the K1, K2 bytes of the line overhead in the first STS-1 in a SONET signal.

This messaging capability is defined in TR-TSY-00253 and in T1.105. There are different definitions for the messages depending on whether the APS is for Rings or for linear SONET networks. T1.105.01-1994 defines the messaging for bi-directional line switched rings.

Path Overhead Bytes

Section overhead / Line overhead	Framing A1	Framing A2	STS-1 ID C1		Path trace J1
Section overhead	BIP-8 B1	Orderwire E1	User F1		BIP-8 B3
	Data com D1	Data com D2	Data com D3		Signal label C2
	Pointer H1	Pointer H2	Pointer H3		Path status G1
	BIP-8 B2	APS K1	APS K2		User channel F2
Line overhead	Data com D4	Data com D5	Data com D6		Multiframe H4
	Data com D7	Data com D8	Data com D9		Growth Z3
	Data com D10	Data com D11	Data com D12		Growth Z4
	Growth Z1	Growth Z2	Orderwire E2		Growth Z5

Path Overhead [carried in SPE]

Path overhead functions provide "end-to-end" monitoring and communications capability. As with the section and line overhead, an 8-bit wide bit-interleaved parity check (B3) is calculated over all bits of the previous SPE. The computed value is placed in the Path Overhead of the following frame.

Alarm and performance information is included as part of the Path Overhead (G1). (These functions are discussed in more detail in the context of in-service testing at the end of this section.)

The structure of the Synchronous Payload Envelope is given via a Path "Signal Label" (C2). This is an 8-bit code value which specifies the SPE structure. 256 different structures are possible. For example, the all "0s" code represents SPE is unequipped (i.e. does not contain any signals).

A fixed length string (64 bytes) is transmitted repeatedly one byte per SPE frame (J1). The 64-byte string can contain any alpha numeric message and is associated with the Path. Continuity of connection to the source of the Path signal can be verified, therefore, at any receiving terminal along the path by simply monitoring this message string.

A generalized multiframe indicator is provided for payloads (H4). At present this facility is used only by VT structured payloads. For example, the VT Overhead and SPE are distributed across 4 frames comprising a VT multiframe. The value of the multiframe indicator in the Path Overhead identifies the phase of the VT multiframe being carried by that SPE.

A User Channel (F2) is provided for proprietary network operator communications between Path Terminating Equipment.

The Tandem connection information (Z5) has recently been adopted to provide Tandem path information for multi-carrier transport of signals.

Path Overhead Functions

- B3—Parity check

- G1—Alarm and performance information
  - −FEBE
  - −Path yellow

- C2—Signal label
  - −Structure of SPE

The nine bytes of STS-1 Path Overhead are carried in the SPE and are made up as follows.

B3: The B3 byte provides a BIP-8 "path" error monitoring function. The path BIP-8 is calculated over all bits of the previous SPE and the computed value is placed in the B3 byte before scrambling.

G1: The G1 byte is used to convey back to the originating STS Path Terminating Equipment the path-terminating status and performance. This feature allows the status and performance of a two-way path to be monitored at either end or at any point along the path.

C2: The C2 byte indicates the construction of the STS SPE by means of a label value assigned from a list of 256 possible values.

Path Overhead Functions

- J1—Path trace
  –Repeating 64 byte message

- H4—Multiframe indication for VTs

- F2—User channel

- Z5—Tandem path performance

J1: The J1 byte is used to repetitively transmit a 64-byte, fixed length string so that continued connection to the source of the path signal can be verified at any receiving terminal along the path.

H4: The H4 byte provides a multiframe phase indication for VT payloads.

F2: The F2 byte is allocated for user's purposes between path termination's, e.g., downloading firmware to terminals.

Z3-Z4: The two bytes Z3 and Z4 are reserved for future use.

Z5: The Z5 byte is used to convey information about the Tandem Connection, including an incoming error count and a tandem connection data link.

H4: The H4 byte provides a multiframe phase indication for VT payloads.

VT Overhead Byte (V5)

- First byte in VT SPE

- BIP-2 (parity)

- VT path FEBE

- VT signal label

- VT path yellow

- Growth

The first byte in the VT SPE is known as the V5 VT overhead byte. This byte provides a:

BIP-2 parity calculation. The parity of the previous VT SPE is calculated across the even bits and odd bits and set in bits 1 and 2.

Path FEBE (Far End Block Error). Bit 3 indicates if any errors were received by the path terminating equipment to convey back to the VT PTE the path performance.

Spare capacity for future use. Bit 4 is currently undefined.

VT signal label. Bits 5-7 provide signal label which identifies the VT mappings in use.

Path Yellow. Bit 8 is set to indicate a path yellow. Industry usage is now moving toward calling this failure a Remote Alarm Indicator (RAI).

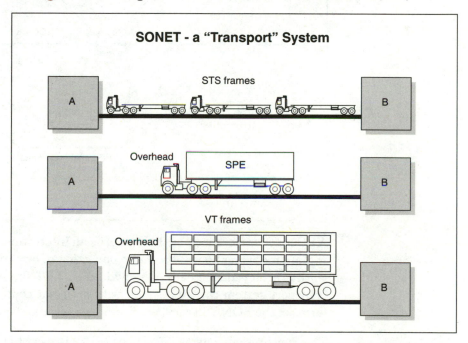

The transport system adopted in SONET is analogous to a road transport system. If you need to deliver items between 2 points you need trucks. Depending on the quantity of items to be moved you need small or large trailers.

Depending on the size of the items being shipped you need "pallets" to allow simple stacking within the trailer payload area. For different item types you have different pallets and different loading instructions. SONET has exactly the same concepts with different names:

• Road —Optical Carrier

• Truck —Synchronous Transport Signal

- Trailer —Synchronous Payload Envelope
- Pallets —Virtual Tributary Frames

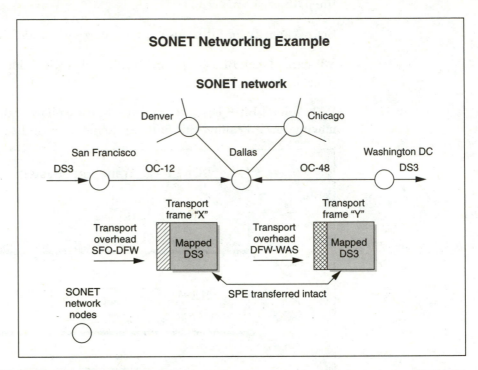

A SONET network may be thought of as an interconnected mesh of SONET signal processing nodes. The interconnection of any two nodes in this network is achieved by means of individual SONET Transport Systems. Each transport system carries a signal with a format that may be described in terms of the SONET Transport Frame Structure.

The Synchronous Payload Envelope (SPE) is used to transport a tributary signal across the synchronous network. In most cases, this signal is assembled at the point of entry to the synchronous network and disassembled at the point of exit from the synchronous network. Within the synchronous network, however, the Synchronous Payload Envelope is passed on intact between transport systems on its route through the network.

Transport Overhead is created on the transmit side of each network node and terminated at the downstream receiving network node. Thus, the Transport Overhead pertains only to an individual transport system and supports the transportation of the SPE over that transport system. It is not transferred with the SPE between transport systems.

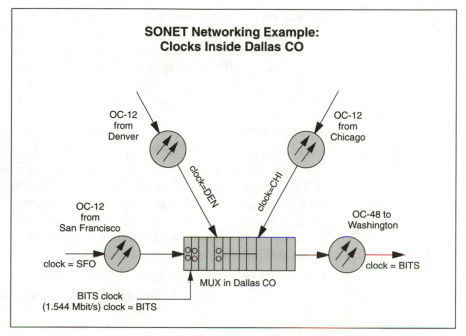

SONET Networking Example:
Clocks Inside Dallas CO

At a typical Central Office, SONET signals from different parts of the network may be multiplexed together. The line rates of these may be traceable back to different reference clocks, either because these SPEs were assembled at another phone company using a different clock source, or because of a synchronization fault. The outgoing line rate is defined by the local BITS clock and the small difference between this timing source and the incoming clock needs to be accommodated as discussed next.

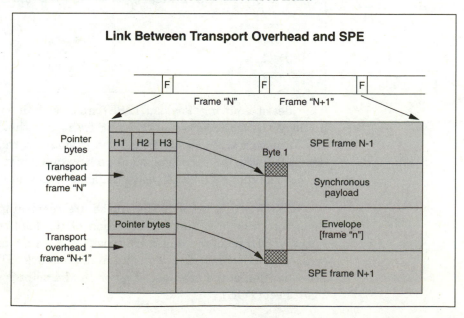

Link Between Transport Overhead and SPE

To take care of small timing differences in the synchronous network, and simplify multiplexing and cross-connection of signals, the SPE is allowed to float within the payload capacity provided by the STS-1 frames. This means that the STS-1 SPE may begin anywhere in the STS-1 payload capacity and is unlikely to be wholly contained in one frame. More likely than not, the STS-1 SPE begins in one frame and ends in the next.

When an SPE is assembled into the Transport Frame, additional bytes, referred to as the "Payload Pointer", are made available in the Transport Overhead. These bytes contain a pointer value which indicates the location of the first byte (byte 1) of the STS-1 SPE. The SPE is allowed to float freely within the space made available for it in the transport frame so that timing phase adjustments can be made as required between the SPE and the transport frame. The payload pointer maintains the accessibility of the SPE by identifying the first byte location of the SPE.

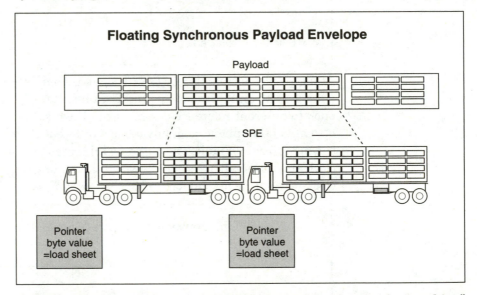

This idea of a moving "Payload" (SPE) can be explained using the "trucking" analogy. The truck represents the STS frame and the drivers loading sheet the Transport Over Head (TOH). The payload on the trucks can be split across 2 trucks the split dependent on how much of the load was available at the time the truck had to leave the loading bay.

As SONET is synchronous the trucks are constrained to an arrival and departure schedule. The start position of the load is given by the "load sheet" (TOH) pointer value. Each truck (STS frame) is always fully loaded with "payload" (SPE) of its own and from the truck (STS frame) before. The start position of the new load is given by the pointer values shown on the load sheet (TOH).

Simplified Drop and Insert

- No need to perform stage-by-stage demultiplexing to recover tributaries

- Pointers give exact position of any customer's data

- STS-1 and VT pointers

As the STS-1 pointers give the exact position of any STS payload that payload can be accessed directly without the need to de-multiplex the SONET line signal. This makes devices like the Add Drop Mux and SONET cross-connect switches much simpler than existing telecom systems requiring full demux before switching. In a similar manner any individual customer data channel can be accessed within an SPE.

VT pointers are used to provide the same capabilities and advantages for VT mapped signals, for example, being able to drop a DS1 directly from a STS-1 SPE with VT1.5 mapping.

What Do Payload Pointers Do?

- Allow asynchronous operation within limited clock offsets

- Minimizes network delay compared with existing multiplexers, e.g., M13

- Introduces a new signal impairment
  −Pointer adjustment jitter

SONET is intended to be a synchronous network. Ideally, this means that all synchronous network nodes should derive their timing signals from a single master network clock.

However, current synchronous network timing scenarios do allow for the existence of more than one master network clock. For example, networks owned by different network operators may have their own independent master timing references. These clocks would operate independently and, therefore, at slightly different rates. Also, the situation in which a network node loses timing reference and operates on a stand-by clock needs to be handled. Synchronous transport must be able, therefore, to operate effectively between network nodes operating asynchronously.

To accommodate clock offsets, the SPE can be moved (justified), positively or negatively one byte at a time, with respect to the transport frame. This is achieved by simply re-calculating or updating the Payload Pointer at each SONET network node. In addition to clock offsets, updating the Payload Pointer will also accommodate any other timing phase adjustments required between the input SONET signals and the timing reference of the SONET node.

Pointers also minimize network delay in the synchronous network. Another approach to overcoming network timing issues is to use 125 ms SPE slip buffers at the inputs to SONET multiplexing equipment to phase align and slip (repeat or delete a frame of information to correct frequency differences) the individual SONET tributary signals as required. These slip buffers are undesirable, however, because of the signal delay that they impose and the signal impairment that slipping causes. Using Payload Pointers avoids these unwanted network characteristics.

Payload Pointer processing does, however, introduce a new signal impairment known as "Payload Output Jitter." This jitter impairment appears on a received tributary signal after recovery from an SPE which has been subjected to Payload Pointer changes. Excessive jitter on a tributary signal will influence the operation of the network equipment processing the tributary signal immediately downstream. Great care is, therefore, required in the design of the timing distribution for the synchronous network. This is in order to minimize the number of Payload Pointer adjustments and, therefore, the level of tributary jitter that could be accumulated through synchronous transport.

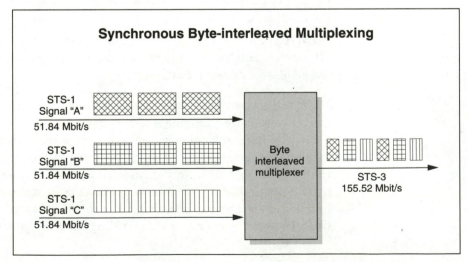

Groups of synchronous transport frames may be packaged for transportation as a higher order synchronous transport signal. Higher order grouping is achieved by means of the process of byte-interleave multiplexing whereby parallel streams of transport signals are mixed together on a fixed byte by byte basis. These parallel streams of transport signals are required to have the same frame structure and bit rate and in addition be frame synchronized with each other.

Thus, for example, 3 parallel and frame synchronized STS-1 SONET signals may be byte-interleaved multiplexed together to form an STS-3 SONET signal at 3 ¥ STS-1 bit rate. Byte-interleaved multiplexing is accomplished by taking, in turn, one byte from STS-1 signal "A" for output, followed by one byte from STS-1 signal "B" for output, followed in turn by one byte from STS-1 signal "C" for output. At this point, the sequence is repeated by returning to STS-1 signal "A" to provide the next byte for output.

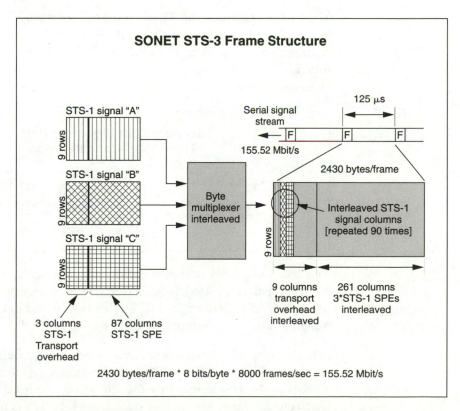

The SONET STS-3 signal is assembled by byte-interleaving 3 parallel frame synchronized STS-1 signals. Consequently, a two dimensional map for the STS-3 signal frame comprises the same 9 row depth as the STS-1

signal but has 270 columns which is 3 times the number of columns of the STS-1 signal. The total signal capacity of the STS-3 signal is, therefore, 2430 8-bit bytes or 19,440 bits per frame.

With these frame dimensions and a frame repetition rate of 8000 frames/s, the signal rate for the STS-3 signal is 155.52 Mbit/s. (Note, the frame repetition rate of a SONET signal is 8000 frames/s irrespective of the level of that signal.)

The 2-dimensional map of the STS-3 signal is assembled by taking individual columns from each of the three STS-1 signal structures and interleaving these in a repeating sequence. Thus, starting with the first columns of each STS-1, one column is taken from STS-1 number 1, followed by one column from STS-1 number 2, followed by one column from STS-1 number 3. This sequence is then repeated 90 times until all the columns are assembled.

The first 9 columns of the STS-3 frame are occupied by the Transport Overhead. These 9 columns are not all used completely. The remaining 261 columns are occupied by the three SPE signals associated with the three individual STS-1 signals. These signals are byte-interleaved by columns as described above.

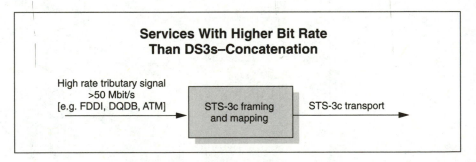

The more advanced customer service signals, such as FDDI, DQDB and ATM (Broadband ISDN), require transport capacity greater than the 49.53 Mbit/s provided by the STS-1 SPE. This is achieved in SONET by means of a higher rate "concatenated" SPE.

A higher rate STS-3 transport signal is normally assembled by byte-interleaving multiplexing three STS-1 transport signals which contain tributary signals at the DS3 signal rate (44.736 Mbit/s) or less. In the SONET context, concatenation means that the higher rate STS-3 transport signal provides effectively one single SPE with a larger payload capacity. A higher rate tributary signal (> 50 Mbit/s) is mapped directly into the larger payload capacity of the STS-3c (where c denotes concatenation) transport signal. The STS-3c SPE is assembled without ever going through the STS-1 signal level.

Once assembled, a concatenated SPE is multiplexed, switched and transported through the network as a single entity.

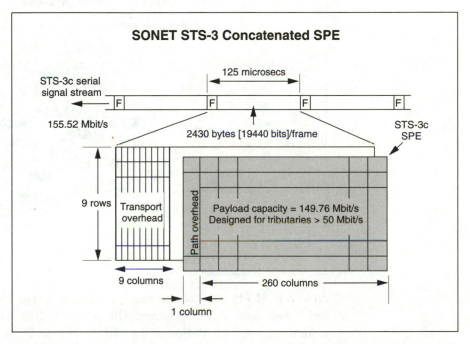

The STS-3c signal frame has the same overall dimensions, 9 rows by 270 columns, the same frame repetition rate, 8000 frames per second, and therefore the same signal rate, 155.52 Mbit/s, as the standard STS-3 signal.

Also in common with the standard STS-3 signal, the first 9 columns of the STS-3c frame, a total of 81 bytes, are allocated to Transport Overhead.

The STS-3c Payload Capacity comprises 260 columns of 9 bytes —a total of 2340 bytes. These bytes provide a transport capacity of 149.76 Mbit/s at a frame repetition rate of 8000 Hz.

Signal capacity for Path Overhead is allocated in the first column of the STS-3c SPE —a total of 9 bytes per frame.

STS-12c is becoming important for supercomputer networking, mapping HPPI and other interfaces into a 600 Mbit/s payload.

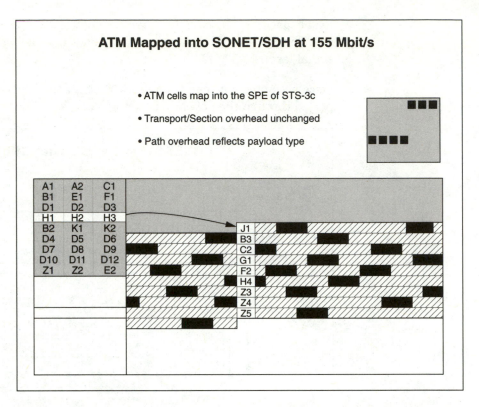

ATM Mapped into SONET/SDH at 155 Mbit/s

- ATM cells map into the SPE of STS-3c
- Transport/Section overhead unchanged
- Path overhead reflects payload type

ATM cells (53 bytes, 5 of overhead, 48 of payload) are mapped directly into the SPE without any delineation. This is accomplished by performing a continuous sliding calculation of the Header Error Correction code on 4 preceding bytes.

When the calculation is correct, a cell header may have been detected, and the ATM receiver goes through hunt and sync states to confirm this. The cell stream remains delineated after this until either the path is broken, or serious HEC errors occur. An older proposal to point to cell boundaries with the H4 byte of the path overhead is now obsolete.

Cell scrambling to an $x43 + 1$ algorithm is done prior to mapping cells into the SPE. This is in addition to SONET scrambling.

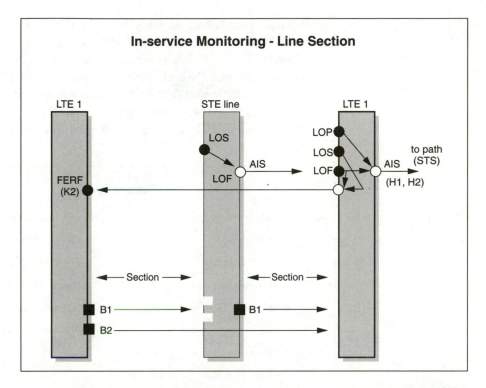

Special signal functions built into particular overhead bytes make effective "in-service" testing possible within a SONET network. Failures such as Loss of Signal (LOS), Loss of Frame (LOF), and Loss of Pointer (LOP) cause Alarm Indication Signal (AIS) to be transmitted downstream. Different AIS signals are generated depending upon which level of the maintenance hierarchy is affected.

In response to the different AIS signals, and failures, other maintenance signals are sent upstream to warn of trouble downstream. Far End Receive Failure (FERF) is sent upstream in the Line Overhead after Line AIS, or LOS, or LOF has been detected by equipment terminating in a line span; a Remote Alarm Indication (RAI) for an STS Path, which is also referred to as STS Path Yellow, is sent upstream after STS Path AIS or STS LOP has been detected by equipment terminating an STS Path.

Similarly, a Remote Alarm Indication (RAI) for a VT Path, which is also referred to as VT Path Yellow, is sent upstream after VT Path AIS or VT LOP has been detected by equipment terminating a VT Path. The path terminating equipment's report to the management OSS the alarm condition.

Performance monitoring at each level in the maintenance hierarchy is based on Bit-Interleaved-Parity (BIP) checks calculated on a frame by frame basis. These BIP checks are inserted in the overhead associated with the Section, Line and Path maintenance spans. In addition, equipment terminating STS Path and VT Path spans produce Far End Block Error (FEBE) signals based on errors detected in the STS Path and VT Path BIPs respectively. The FEBE signals are sent upstream to the originating end of a Path.

Summary

SONET technology

- STS-1

- SPE mapping

- Payload mapping

- Byte interleaving, concatenation

Benefits to the SONET user

- Vendor compatibility

- Fast provisioning

- Bandwidth management

- Network survivability

- Potential for new services

Alarms and fault reporting, Control

- Path, Line, Section Overheads

- User channel, DCC

Let's now review the ideas and concepts covered regarding SONET.

What is SONET defined to be?
-a standard for a high-speed optical-fiber-based network.

What devices have been defined as network elements?
-DLC, TM, ADM, DCS and regenerators.

Who has defined the SONET standards?
-ANSI, Bellcore, CCITT.

The theory and operation of the SONET technology?
-OC-n, STS-n, SPE, VTs, Superframes, Concatenation.

The services to be transported by SONET?
-DS0, DS1, DS3, HDTV, BISDN, ATM.

The resilience of the SONET network due to "overhead" providing status, alarm and management information?
-POH, LOH, SOH, management data communication channels, standardized maintenance messages.

The benefits to network operators in changing their transmission networks to SONET?
-commodity network element market due to standards.

Faster and better network management and control?
-provisioning, service re-routing, billing, bandwidth utilization, fault isolation.

Potential for new bandwidth hungry services?
-LAN, MAN, WAN interconnects,
BISDN, CAD/CAM, Video on demand.

Testing SONET Networks

CERJAC Telecom Operation
43 Nagog Park
Acton, MA 01720

SONET/ATM
Networks and Testing Seminar

Abstract

SONET has moved from the R&D labs into the field, bringing with it a host of new installation and maintenance test requirements. The SONET standard allows network equipment from different vendors to be connected at an optical interface point. This compatibility at high rate interfaces enables greater flexibility in configuration of complex networks incorporating rings, linear add-drop systems, and broadband digital cross-connect systems. Testing these new network topologies requires test equipment which can interface at the SONET standard rates.

As SONET deployment continues, new test and maintenance issues will continue to emerge. SONET networks of the future will feature multiple rings, interfaces among SONET network elements at many different OC-N rates, and complex tributary mapping and routing plans. The ability to routinely access and test at these OC-N interfaces, and to map from any one interface rate to any other will be necessary for network implementers and maintenance personnel. And as SONET network complexity increases, so will the requirements for sophisticated test and diagnostic procedures.

This paper discusses situations which can arise in the deployment of SONET technology, and presents test methods to address these new challenges.

Author

CERJAC Telecom Operation
43 Nagog Park
Acton, MA 01720

Hewlett-Packard

Testing SONET Networks

SONET has moved from the R&D labs into the field, bringing with it a host of new installation and maintenance test requirements. This paper discusses situations which can arise in the deployment of SONET technology, and presents test methods to address these new challenges.

Seminar Content

- SONET testing background

- Out-of-service equipment tests

- Network turn-up and service verification

- Alarm and APS testing

- Testing Rings

- Synchronization tests

- Summary

Background to Asynchronous Testing

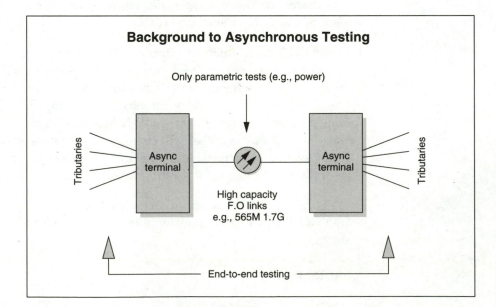

Prior to the deployment of SONET, high speed communication was accomplished using asynchronous fiber-optic transmission systems. Since there was no standard for the optical interface, these systems employed proprietary optical rates and formats.

Testing of these asynchronous systems is carried out end-to-end on a complete line system, usually at a DS3 or DS1 tributary interface. Each tributary channel of the line system is stimulated on one side by means of a test signal incorporating a Pseudo Random Binary Sequence (PRBS). On the other side, errors in the PRBS bit pattern are detected and a Bit-Error Ratio (BER) measurement is made as a measure of performance.

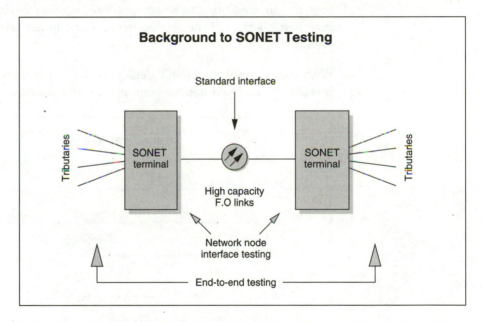

Now, however, with defined standards for SONET, network equipment from different vendors can be connected at the optical interface point. This compatibility at high rate interfaces also enables greater flexibility in configuration of complex networks incorporating rings, linear add-drop systems, and broadband digital cross-connect systems.

Testing these new network topologies requires SONET-compatible test equipment which can interface at the SONET standard rates.

Installation and Maintenance Testing

- SONET equipment is now routinely installed for network growth.

- Services delivered over SONET include DS0, DS1, DS3, OC-3c and ATM.

- SONET interfaces are provided for end users and at inter-network POPs.

- Installation and maintenance of SONET networks is now a high priority
  revenue-critical activity.

Virtually all major carriers now deploy SONET equipment for network growth. In addition, new competitive access carriers have typically installed all-SONET networks, bypassing the asynchronous stage completely.

With this increasing SONET deployment has come an increased need for installation and maintenance equipment and procedures.

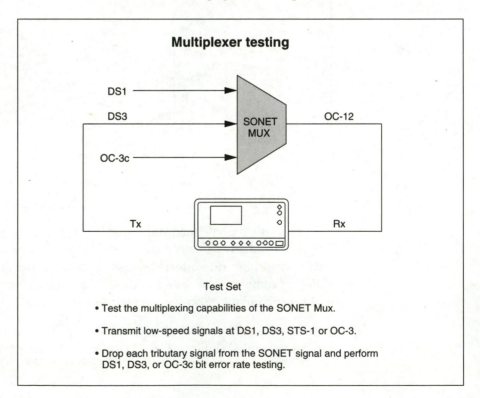

Multiplexer testing

DS1

DS3   SONET   OC-12
             MUX

OC-3c

Tx   Rx

Test Set

- Test the multiplexing capabilities of the SONET Mux.

- Transmit low-speed signals at DS1, DS3, STS-1 or OC-3.

- Drop each tributary signal from the SONET signal and perform
  DS1, DS3, or OC-3c bit error rate testing.

Pre-installation testing is sometimes performed on network elements. For example, the performance of the payload mapping process in a SONET multiplexer can be checked.

The SONET tester provides a test stimulus, such as a DS3, DS1, or SONET STS-1, OC-3, or OC-3c signal, normally containing a PRBS running at the desired bit rate. After the test tributary signal has been mapped into a SONET signal by the multiplexer, the SONET tester recovers the test tributary for analysis.

The mapping process is verified by proving the integrity of the recovered tributary signal with a BER test. This requires that a tributary demapping capability must be provided in the SONET tester.

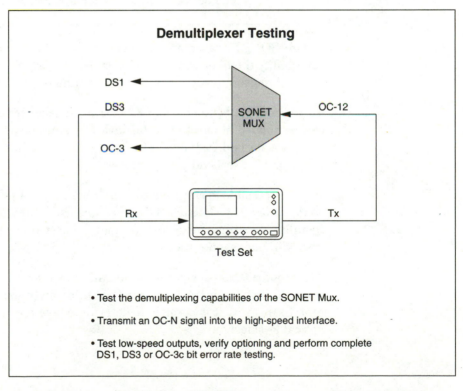

Demultiplexer Testing

- Test the demultiplexing capabilities of the SONET Mux.

- Transmit an OC-N signal into the high-speed interface.

- Test low-speed outputs, verify optioning and perform complete DS1, DS3 or OC-3c bit error rate testing.

For demultiplexer testing, the SONET tester maps a tributary signal into a SONET structure and applies it to the high-speed SONET interface. The SONET multiplexer under test demaps the tributary signal and makes it available at an appropriate tributary interface. The tester is then connected to the tributary output port to measure the BER and verify the integrity of the tributary signal.

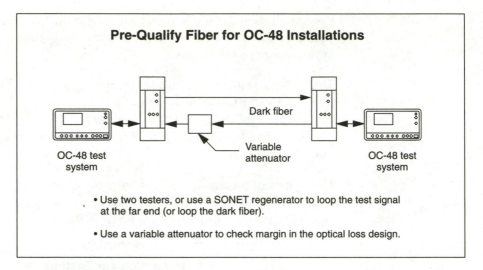

Pre-Qualify Fiber for OC-48 Installations

OC-48 test system

Dark fiber

Variable attenuator

OC-48 test system

- Use two testers, or use a SONET regenerator to loop the test signal at the far end (or loop the dark fiber).

- Use a variable attenuator to check margin in the optical loss design.

Before SONET transport systems are turned-up, it is often necessary to pre-qualify the fiber facilities. This is particularly true when existing fiber is being re-used to support higher rates than originally deployed.

To pre-qualify fiber spans for OC-48 transmission, for example, a test signal can be applied at one end of the dark fiber, and measured by a second test set at the far end. Extended tests can be run, monitoring payload bit errors and SONET overhead parity errors.

It is also possible to employ a SONET regenerator to loop the far-end signal so that a single OC-48 test system can be used. In relatively short distance installations, it may also be reasonable to provide a direct loop of the dark fiber without regeneration.

After confirmation of error-free transmission at nominal design levels, additional loss can be introduced into the optical path using a variable attenuator. In this way it is possible to determine how much margin exists in the optical loss design.

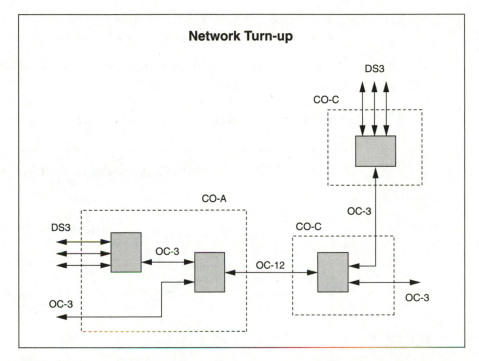

SONET terminals often include diagnostics to facilitate the installation of simple point-to-point configurations. More complex networks, however, are best tested segment by segment before final end-to-end testing is attempted. In addition, as shown above, it may be necessary to turn up and verify each central office separately before connecting the fiber spans.

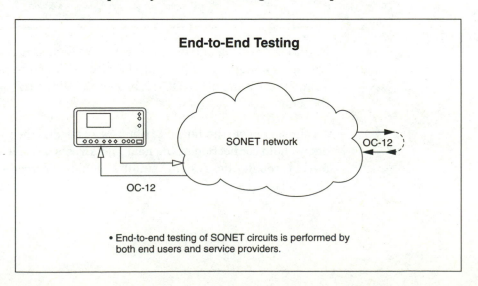

The aim of end-to-end testing is to verify that the integrity of an STS or VT payload is maintained through the network under test.

On the transmit side of the tester, a PRBS is loaded into the selected SPE in a structured SONET signal. This test stimulus is transmitted through the network. At the receive side of the tester, the test SPE is demultiplexed and its integrity verified by a BER measurement on the recovered PRBS.

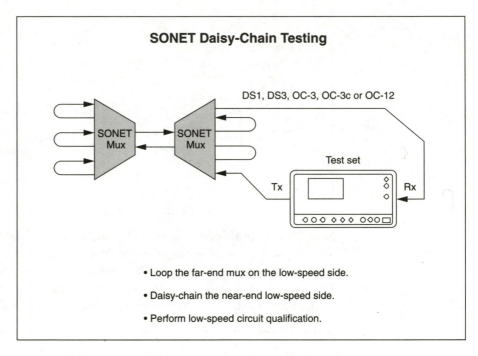

SONET Daisy-Chain Testing

DS1, DS3, OC-3, OC-3c or OC-12

SONET Mux

SONET Mux

Test set

Tx   Rx

- Loop the far-end mux on the low-speed side.
- Daisy-chain the near-end low-speed side.
- Perform low-speed circuit qualification.

End-to-end testing of tributaries can be extended to multiple channels of a transport system through daisy-chaining. This can be done with tributaries at any rate, from DS1 to OC-12, as long as all the tributary channels are at the same rate.

As shown above, the far end tributary ports are each looped back, while a daisy chain connection at the near end directs a single test signal back and forth through the system, simultaneously exercising all the tributary channels.

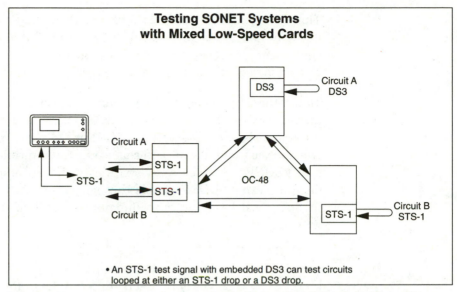

**Testing SONET Systems
with Mixed Low-Speed Cards**

• An STS-1 test signal with embedded DS3 can test circuits looped at either an STS-1 drop or a DS3 drop.

OC-48 transport systems frequently employ either DS3 or STS-1 tributary interfaces. Increasingly, network topologies are being designed which require individual tributaries on OC-48 systems to have STS-1 interfaces on one end and DS3 interfaces on the other end. This facilitates connection to DS3 distribution points on one end, and extension of the circuit via STS-1 to other SONET network elements on the other end.

In these cases, a SONET tester providing an STS-1 signal with an embedded DS3 payload can be used to test either circuit type. The test signal is applied at an STS-1 interface. The far-end signal is then looped back at either STS-1 or DS3.

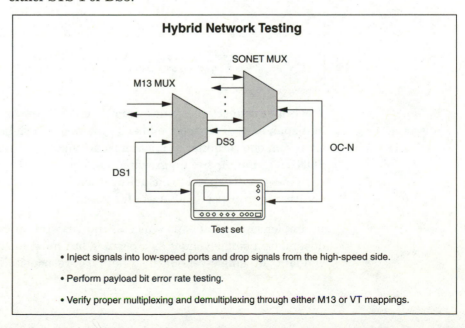

Hybrid Network Testing

• Inject signals into low-speed ports and drop signals from the high-speed side.

• Perform payload bit error rate testing.

• Verify proper multiplexing and demultiplexing through either M13 or VT mappings.

There is an extensive embedded base of DS1/DS2/DS3 (M13) multiplexers in the network. And, although VT1.5 (Virtual Tributary, 1.5 Mbit/s) based equipment is now being deployed, a large number of DS1s will continue to be transported within DS3s over the SONET network.

To test these hybrid network configurations, it is necessary to generate and measure hybrid T-carrier and SONET signals. This requires the ability to inject signals at a low-speed port (DS1 or DS3) and measure the resulting transmission and alarm performance in the multiplexed outputs at high speed SONET ports, or vice versa. Verification of proper multiplexing and demultiplexing of DS1 signals through either M13 or Virtual Tributary mapping will be a routine maintenance requirement as SONET islands grow.

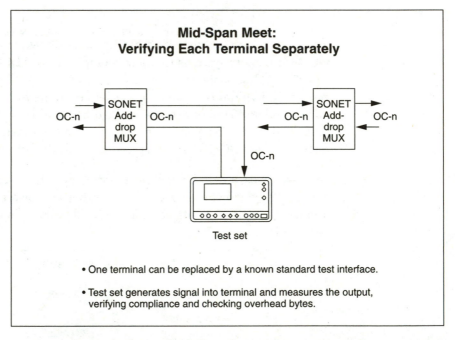

**Mid-Span Meet:
Verifying Each Terminal Separately**

Test set

• One terminal can be replaced by a known standard test interface.

• Test set generates signal into terminal and measures the output, verifying compliance and checking overhead bytes.

Fiber-optic connection of equipment from different vendors is known as a mid-span meet. Transport mid-span meet, in which basic signal transmission from one terminal to another is achieved, is already a reality among SONET terminals from a growing number of vendors. In the future, mid-span meets will expand to include overhead communications capabilities, in addition to transport and alarm functions.

In implementing mid-span meets, the network provider can no longer depend on a single vendor as a partner, but must mediate among multiple vendors. In single-vendor installations, responsibility for problems at installation, turn-up or during operation can be placed clearly with the equipment vendor. In multi-vendor environments, however, the network provider must have the tools and procedures necessary to identify which vendor's equipment is at fault should problems arise.

This requires an external test capability to serve as a "referee" and perform the mediation function. Each terminal in turn can be replaced by this known standard test interface. The test set then generates SONET signals into the terminal, and measures the terminal's output, verifying compliance and operation of the equipment. In this way, a failed network element can be isolated.

International Gateways: Check the H1 Byte!

- European (SDH) systems set bits 5&6 of the H1 pointer byte to 01.

- Some US equipment can't tolerate this value.

- New international trunks to the US often follow the SDH standard.

- Check the H1 byte value when experiencing difficulties with SONET equipment interfacing at these international gateways.

New international fiber installations are adopting SDH (Synchronous Digital Hierarchy) standards. While similar in most respects to SONET, the SDH signal asserts a 01 value in bits 5 and 6 of the H1 pointer byte. The SONET standard identifies these bits as "don't care." This means that SONET systems should be able to tolerate any value in these bits. In practice, however, SONET equipment may not accept these signals, resulting in alarm or failure conditions.

It is useful to check the H1 byte on signals at international gateways before connection to domestic SONET equipment.

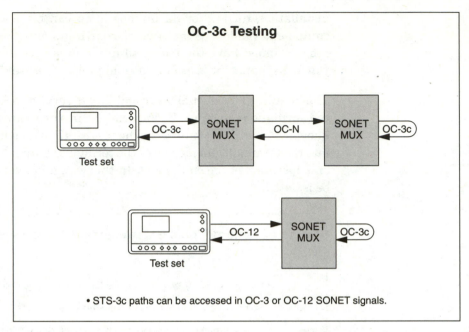

OC-3c Testing

• STS-3c paths can be accessed in OC-3 or OC-12 SONET signals.

OC-3 transport of ATM signals makes use of the bandwidth of 3 STS-1 signals concatenated into a single STS-3c payload. Clear channel STS-3c testing is required to pre-qualify transmission links before ATM traffic is applied. The STS-3c payload is transmitted within either an OC-3 or higher-rate OC-N signal.

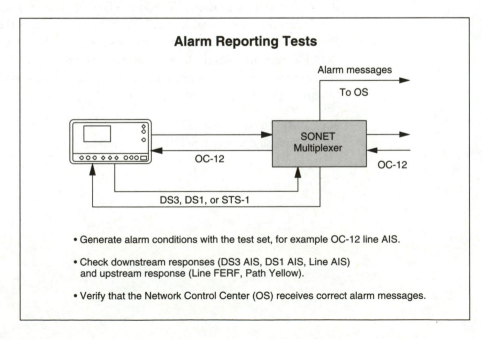

Alarm Reporting Tests

• Generate alarm conditions with the test set, for example OC-12 line AIS.

• Check downstream responses (DS3 AIS, DS1 AIS, Line AIS) and upstream response (Line FERF, Path Yellow).

• Verify that the Network Control Center (OS) receives correct alarm messages.

Alarm detection and processing can be tested through controlled external generation and detection of alarm conditions in the SONET and tributary signals.

For each category of SONET equipment, specific up-stream and down-stream alarm responses are required. On detection of an incoming alarm condition such as Loss of Signal (LOS), or Loss of Frame (LOF), the equipment must send appropriate indications toward the trouble signal (up-stream) and away from the trouble signal (down-stream). External test sets must provide the ability to force these alarm conditions, and to detect the alarm responses received from the unit under test.

In many cases, SONET terminals are connected to Operations Systems for reporting of alarm conditions. It is advantageous to verify that this alarm communication is working properly. Once again, the external test set's ability to force alarm conditions is used. The required alarm conditions can be forced, and the reports to the OS can be verified.

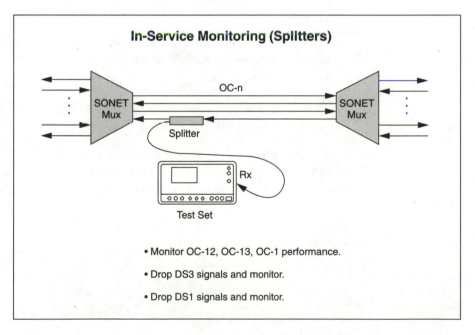

Optical splitters can be used to gain access to an optical SONET signal. Alternatively, a test set pass-through feature can be used to regenerate the down-stream SONET signal. In either case, extreme care must be used to avoid inadvertent service interruption.

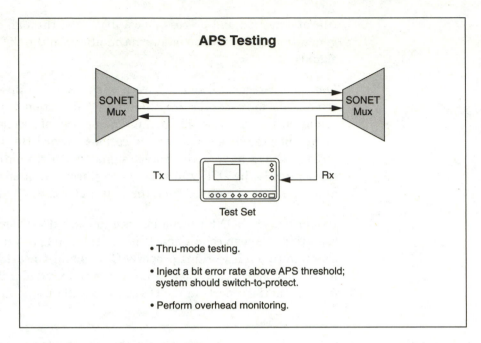

With a test set configured for through-mode testing, B2 Line Code Violations or line-level SONET alarms can be added to the signal forcing a switch to protection.

This tests point-to-point systems using line switching, and can also be applied to line-switched ring applications.

The Automatic Protection Switching is activated if line BIP error thresholds are exceeded. The error threshold levels are tested by injecting error rates slightly below and slightly above the nominal limit.

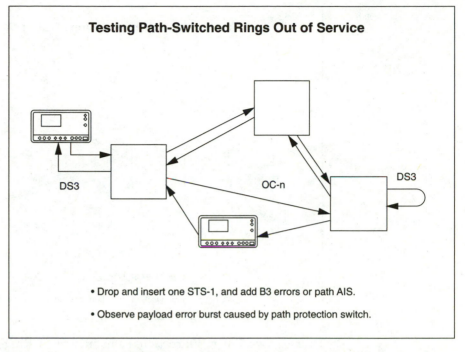

Testing Path-Switched Rings Out of Service

DS3

OC-n

DS3

• Drop and insert one STS-1, and add B3 errors or path AIS.

• Observe payload error burst caused by path protection switch.

In Unidirectional Path-Switched Rings, the decision to switch to protection is made independently for each path on the ring. This means that at any given time, certain paths may be using the main ring, while others are using the protection ring.

For out of service path-switched ring testing, a test set configured for SONET Drop & Insert can be placed in the optical line. B3 Path Code Violations or Path AIS alarm can be added to a selected STS-1 within the OC-3 or OC-12 signal. These violations should then be detected at the path terminating device, causing a path protection switch.

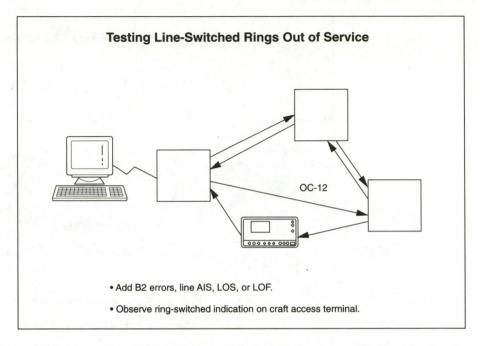

Testing Line-Switched Rings Out of Service

OC-12

• Add B2 errors, line AIS, LOS, or LOF.

• Observe ring-switched indication on craft access terminal.

Four or Two Fiber Bidirectional Line-switched Rings are very similar to point-to-point systems in that they switch based on line parameters (B2 errors and line alarms). For out-of-service testing, a test set in through mode can add these error and alarm conditions, forcing the ring to switch.

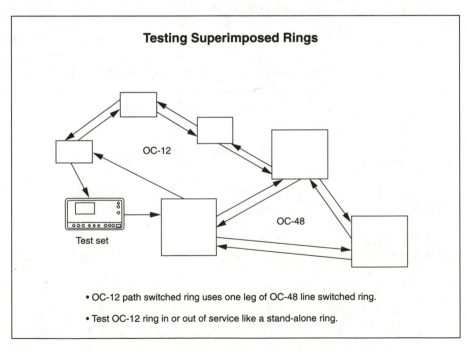

Testing Superimposed Rings

OC-12

OC-48

Test set

• OC-12 path switched ring uses one leg of OC-48 line switched ring.

• Test OC-12 ring in or out of service like a stand-alone ring.

Many networks are now employing multiple interconnected rings. For example, it is not uncommon for an OC-3 or OC-12 path switched ring to have one or more of its spans actually transported over a higher rate line-switched ring. In these cases, the lower rate ring can be accessed on one of its native spans just like a stand-alone path switched ring.

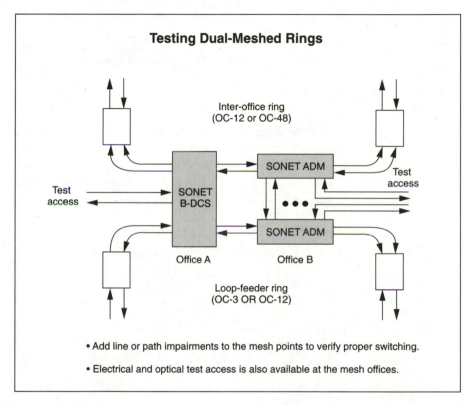

Testing Dual-Meshed Rings

Inter-office ring
(OC-12 or OC-48)

SONET ADM

Test
access

Test
access

SONET
B-DCS

SONET ADM

Office A

Office B

Loop-feeder ring
(OC-3 OR OC-12)

• Add line or path impairments to the mesh points to verify proper switching.

• Electrical and optical test access is also available at the mesh offices.

Another ring inter-working configuration involves dual-meshed rings. Traffic from one ring to another is guaranteed even if one of the mesh offices goes out of service. This functionality can be tested by inserting impairments at the line or path level in one of the mesh interconnections. The traffic should switch to the other connection with no more than a protection error burst.

Ring mesh points also provide circuit access opportunities, through the test facilities of a Broadband Cross-connect System, or through low-speed ports on interconnected Add-Drop Multiplexers.

Timing and Synchronization

- Clock differences between SONET signals are compensated for by pointer adjustments.

- Each pointer adjustment results in a positive or negative "stuff" of 8 bits, resulting in tributary jitter.

- To identify timing problems:
  - Measure clock frequencies recovered from SONET signals.
  - Count SONET pointer adjustments.
  - Measure DS1 timing slips.
  - Measure tributary jitter.

Timing distribution is critical in SONET networks. In the installation and maintenance of SONET systems, it is therefore necessary to identify and correct faulty or drifting timing sources. Test methods can range from measurement of derived SONET clock signals to monitoring pointer adjustment rates or measuring tributary jitter.

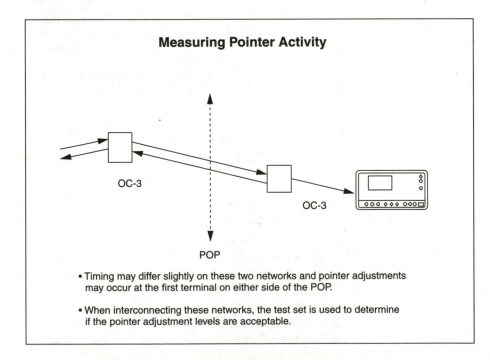

Measuring Pointer Activity

OC-3

OC-3

POP

- Timing may differ slightly on these two networks and pointer adjustments may occur at the first terminal on either side of the POP.

- When interconnecting these networks, the test set is used to determine if the pointer adjustment levels are acceptable.

The preceding illustration depicts an OC-3 connection between two networks at a Point-of-Presence. What is of greatest interest in connecting these two networks is the level of pointer activity caused by their interconnection. Since their clocks may be slightly different, pointer adjustments may occur at the first terminal on either side of the POP. In order to measure these adjustments, the OC-3 signal is monitored just beyond the first SONET multiplexer on either side of the network interconnection point.

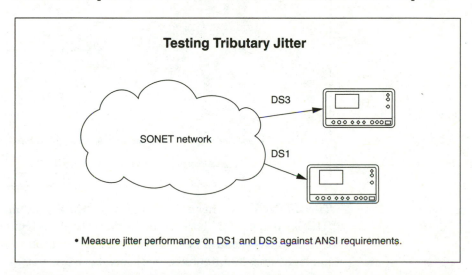

Testing Tributary Jitter

SONET network

DS3

DS1

• Measure jitter performance on DS1 and DS3 against ANSI requirements.

SONET pointer adjustments can cause jitter on tributary DS3 or DS1 signals. The jitter accumulates as the tributary signal crosses through multiple SONET islands, with potential impact on clock synchronization and error performance in the receiving terminals.

This makes tributary jitter measurement a must for networks carrying DS3 or DS1 services over SONET networks. Tributary jitter measurements allow the service provider to verify jitter performance against industry standard masks for DS1, E1, DS3, and STS-1 signals.

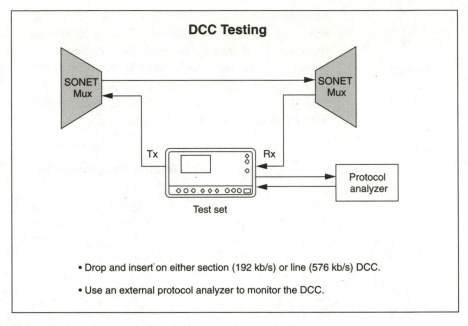

DCC Testing

• Drop and insert on either section (192 kb/s) or line (576 kb/s) DCC.

• Use an external protocol analyzer to monitor the DCC.

The SONET overhead contains two Data Communication Channels (DCCs), one at 192 kbit/s in the Section Overhead and one at 576 kbit/s on the Line Overhead. These DCCs are used to communicate network management and maintenance messages between network elements and the operations support computer system. The message-based information is transferred by means of an OSI protocol stack running at high speed.

The SONET Tester can provide access to either of the DCCs for an external Protocol Analyzer so that the test response messages may be isolated and analyzed.

Summary

• SONET installation and maintenance activity is increasing.

• Some SONET tests are similar to asynchronous methods.

• There are also new requirements like Alarm, APS, Ring and Pointer testing.

• The basic test scenarios discussed today can be applied in many SONET installation and maintenance test situations.

As SONET deployment continues, new test and maintenance issues will continue to emerge. SONET networks of the future will feature multiple rings, interfaces among SONET network elements at many different OC-N rates, and complex tributary mapping and routing plans. The ability to routinely access and test at these OC-N interfaces, and to map from any one interface rate to any other will be necessary for network implementers and maintenance personnel. And as SONET network complexity increases, so will the requirements for sophisticated test and diagnostic procedures.

SONET test challenges are evolving from compliance to installation and maintenance. An understanding of the changing installation and maintenance test requirements will assist in the rapid and efficient deployment of new revenue-generating SONET networks and services.

APPENDIX D

Coaxial Cable System Powering

A section of cable plant is shown below with resistances of cable sections shown and the distances between amplifiers noted.

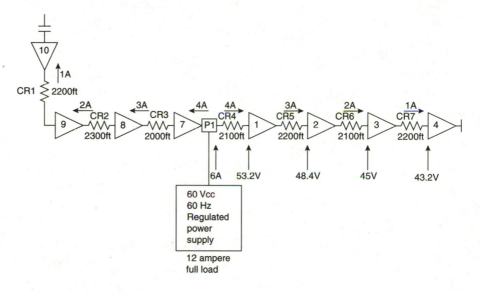

—|⊢— Power block signal pass.

Cable section	Length 1000 ft	Resistance ohms*	Current in amps	Voltage drop † volts †
CR1	2.2	1.8	1A	1.8
CR2	2.3	1.8	2A	3.6
CR3	2.0	1.6	3A	4.8
CR4	2.1	1.7	4A	6.8
CR5	2.2	1.8	3A	4.8
CR6	2.1	1.7	2A	3.4
CR7	2.2	1.8	1A	1.8

Cable loop resistance is 0.8 ohms per thousand feet

The amplifiers each draw 1 ampere each and minimum allowable amplifier voltage is 40 volts

Power supply delivers 8 amps

* 0.8 Ω/1000 [1] × length in thousands of feet.
‡ column 4 × column 3

Now the voltage at each amplifier location can be calculated and labeled on diagram.

Amp 1 60V–6.8 = 53.2V   amp 2 53.2–4.8 = 48.4V   amp 3 48.4–3.4 = 45

Amp 4 45V–1.8 = 43.2V   amp 7 no voltage drop.

Amp 8 60V–4.8 = 55.2V   amp 9 55.2–3.6V = 51.6V   amp 10 51.6–1.8 = 49

All amplifiers have proper voltage, and power supply is only 75% fully loaded.

Broadband Noise Combining

Noise-combining effect in single cable
reverse cable television systems

For a subsplit reverse system where carrier C_1 is at channel T-7 (7.0 MHz) and C_2 is at channel T-8 (13 MHz). And these channels are combined in a splitter and a directional coupler with equal noise levels.

Channel T-7 C_1 = +20 dBmV carrier level N_1 = –25 dBmV noise level

Channel T-8 C_2 = +20 dBmV carrier level N_1 = –25 dBmV noise level

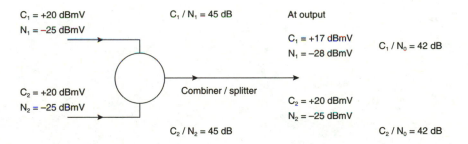

Since noise level is constant across the frequency band the output noise No = –28 dBmV + 3 dBmV = –25 dBmV.

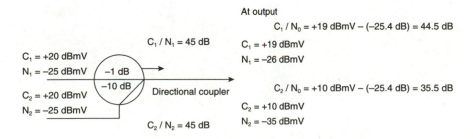

To combine N_1 and N_2 at the output on a power basis

$-26 = 10 \log n_1$	$\log n_1 = -2.6$	$n_1 = 0.00256$
$-26 = 10 \log n_2$	$\log n_2 = -3.5$	$\underline{n_2 = 0.00032}$
	$\log n_0 = n_0$	$n_0 = 0.00288$

$$N_0 = 10 \log n_0 = 10 \log 0.00288 = -25.4 \text{ dB}$$

Summary: In the case of the 2-port combiner, carrier-to-noise level for both C_1 and C_2 was decreased by 3 dB. For the case of the -10 dB directional combiner where C_2 is combined with C_1 through the -10 dB port, C_1 only had a ½ dB decrease in carrier-to-noise level, while C_2 had a 9½ dB decrease in carrier-to-noise level.

APPENDIX F

Cascaded Amplifier Theory

Noise voltage generated by a 75-Ω resistor at 20°C, expressed in dBmV. This voltage appears across the input resistance of the 1st 75-Ω amplifier in a cascade of repeater amplifiers.

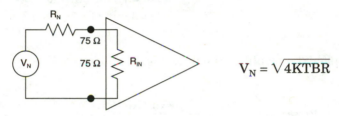

$$V_N = \sqrt{4KTBR}$$

K = Boltzman's constant 1.38×10^{-23}

T = Temperature °K = 20°C + 273°C = 293°K

B = Frequency bandwidth in Hz = 4×10^6 Hz for video bandwidth of 4 MHz

R_N = Resistance in ohms

$$V_N = \sqrt{4 \times 293 \times 1.38 \times 10^{-23} \times 4 \times 10^6 \times 75}$$

$$= \sqrt{485208 \times 10^{-17}} = \sqrt{4.85 \times 10^{-12}} = 2.2 \times 10^{-6} \, V$$

$$= 2/\mu V$$

This is the noise V_N generated by R_N and appears across a circuit of 2 75-Ω resistors connected in series.

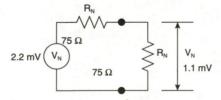

V_{IN} is that portion of noise voltage that appears across the amplifier input terminal and is ½ V_N according to the voltage divider principle.

$$\text{In dBmV} = 20 \log \frac{1.1 \times 10^{-6}\,\text{V}}{1 \times 10^{-3}\,\text{V}} = 20 \log 1.1 \times 10^{-3}$$

$$= 20 \times -2.959 = -59.2 = -59\,\text{dBmV}.$$

This level in dBmV constitutes the so-called noise floor in cable television amplifier cascades.

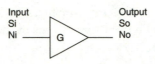

Si = Signal input power level
Ni = Noise input power level
So = Signal output power level
No = Noise output power level
G = Gain of amplifier

For the amplifier the noise factor is given as $F = \dfrac{\dfrac{Si}{Ni}}{\dfrac{So}{No}} = \dfrac{\text{Input signal-to-noise ratio}}{\text{Output signal-to-noise ratio}}$

Rewriting $\qquad F = \dfrac{Si}{Ni} \times \dfrac{No}{So} \left(\dfrac{No}{Ni}\right)\left(\dfrac{Si}{So}\right)$

Since $\qquad G = \dfrac{So}{Si}$ then $\dfrac{Si}{So} = \dfrac{1}{G}$

Eq A substituting $\qquad F = \dfrac{No}{Ni}\left(\dfrac{1}{G}\right)$

The noise power output of an amplifier consists of the input noise level amplified G times, plus the noise Nn generated by the amplifier itself.

Eq B mathematically $\quad No = Ni(G) + Nn$

Eq C combining Eq A and B $\quad F = \dfrac{Ni(G) + Nn}{Ni(G)} = 1 + \dfrac{Nn}{Ni(G)}$

Consider a cascade of 3 amplifiers 1, 2 and 3

Ni → ▷ G_1 G_{N1} F_1 — ▷ G_2 G_{N2} F_2 — ▷ G_3 G_{N3} F_3 → No

For the cascade $\qquad F_C = \dfrac{No}{Ni(G_C)}$ $\qquad$ Gc gain of cascade

No noise power output can be calculated

Eq D $No = Ni\,(G_1)\,(G_2)\,(G_3) + Nn_1\,(G_1)\,(G_2) + Nn_2\,(G_3) + Nn_3$

Cascade input noise	Noise generated by amplifier 1	Noise generated by amplifier 2	3rd amplify. noise

Eq E also $Gc = (G_1)\,(G_2)\,(G_3)$
For the cascade

Eq F substituting Eqc in A $F = \dfrac{Ni\,(G_1)\,(G_2)\,(G_3) + Nn_1\,(G_2)\,(G_3) + Nn_3}{Ni\,(G_1)\,(G_2)\,(G_3)}$

Rewriting Eq F $F = 1 + \dfrac{Nn_1}{NiG_1} + \dfrac{Nn_2}{Ni(G_1)(G_3)} + \dfrac{Nn_3}{Ni(G_1)(G_2)(G_3)}$

$\uparrow$

$F_1\,(Eqc)$

and $F_2 = 1 + \dfrac{Nn_2}{NiG_2}$ $F_3 = 1 + \dfrac{Nn_3}{NiG_3}$

Rearranging the above for F_1, F_2, and F_3

$F_1 - 1 = \dfrac{Nn_1}{NiGi}$ and $F_2 - 1 = \dfrac{Nn_2}{Ni(G_2)}$ and $F_3 - 1 = \dfrac{Nn_2}{NiG_3}$

Substituting back in Eq F.

Eq G for the cascade $F = F_1 + \dfrac{F_3 - 1}{G_1} + \dfrac{F_2 - 1}{G_1 G_2}$ Eq H F in dB = 10 log F
F in dB is the noise figure

Therefore Eq G for M stages becomes:

$$F = F_1 + \dfrac{F_2 - 1}{G_1} + \dfrac{F_3 - 1}{(G_1)\,(G_2)} + \ldots + \dfrac{F_n - 1}{G_1\,G_2\ldots Gn - 1}$$

Given A 2
Amplifier cascade connected by a piece of cable with its loss equal to an amplifier's gain. The cable will generate noise and has negative gain.

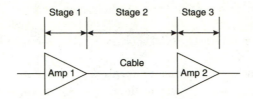

$$F_1 = 10 \text{ dB} \qquad F_2 = 10 \text{ dB} \qquad F_3 = 10 \text{ dB}$$
$$G_1 = 10 \text{ dB} \qquad G_2 = -10 \text{ dB} \qquad G_3 = 10 \text{ dB}$$

Typically, F & G are given in dB form, therefore to use Eq G F & G have to be converted to power ratios.

For amplifier 1 $10 = 10 \log F_1$, $\log F_1 = 1$, $F_1 = \log^{-1} 1$, $F_1 = 10$
                                 in like manner $G_1 = 10$

For the cable $F_2 = 10$ $-10 = 10\log G_2$ $\log G_2 = -1$ $G_2 = \log^{-1}(-1)$
                                   $G_2 = 0.1$

For amplifier 2 $F_2 = 10$ and $G_2 = 10$ (same as amp 1)

Substituting $F = F_1 + \dfrac{F_2 - 1}{G_1} + \dfrac{F_3 - 1}{G_1 G_2}$

$$Fc = 10 + \frac{10 - 1}{10} + \frac{10 - 1}{(10)(0.1)} = 10 + 0.9 + 9 = 19.9 = 20$$

$FcdB = 10\log Fc = 10\log 20 = 10(1.3) = 13 \text{ dB}$

Rule: Doubling number of amplifiers increases noise by 3 dB.

This rule can be applied to a case of many amplifiers (32) in cascade. Since the output signal level is constant and the buildup of noise is increased by 3 dB every time the number of amplifiers is doubled, the signal-to-noise ratio expressed in dB decreases by 3 dB every time the number of amplifiers is doubled.

Since the signals carried on an amplifier cascade are television carriers, for a C/N at amplifier 1 output of 60 dB, consider the example below.

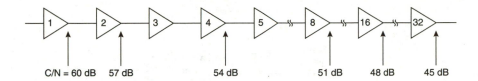

C/N = 60 dB   57 dB   54 dB   51 dB   48 dB   45 dB

The carrier-to-noise ratio is degraded by 3 dB for each time the amplifier number in the cascade is doubled.

45 dB C/N for present-day cable systems is too low. A C/N of 49 dB is acceptable.

To find the C/N of a single amplifier, a test can be performed to find this value. But first consider that for television carrier signal Ci input level & Co output level.

Noise factor $\qquad F = \dfrac{Ci}{Ni} \times \dfrac{No}{Co}$

Rewriting in terms of the output carrier to noise level $\dfrac{Co}{No}$

$$\frac{Co}{No} \times F = \frac{Ci}{Ni} \times \frac{No}{Co} \times \frac{Co}{No}$$

$\qquad\qquad \dfrac{CoF}{No} = \dfrac{Ci}{Ni} \qquad$ Dividing both sides by F

Eq 1 $\qquad \dfrac{Co}{No} = \dfrac{Ci}{NiF}$

Rewriting Eq 6 in terms of dB.

$$\left. ^{Co}\!/\!_{No} \right|_{dB} = Ci_{dB} - (Ni_{dB} + F_{dB})$$

$$= \text{signal input level (dBmV)} - (-59 \text{ dBmV}) - F_{dB}$$

$$\text{C/N} \left|_{\substack{dB}} \right. = 59 - F_{dB} + \text{signal input dBmV}$$
$$\text{dB signal amplifier}$$

The test configuration for the noise figure of an amplifier under test is shown below.

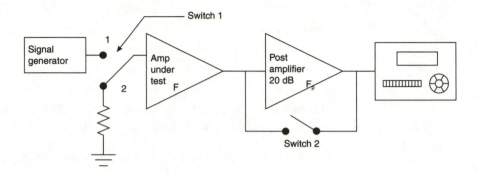

With switch 1 in position 1 and switch 2 closed, gain of amplifier is measured as $\dfrac{Co}{Cin} = gain$

Signal level meter with noise test feature will not need any correction factor.

1) Or Co/dB – Cin/dB = gain/dB = 22 dB. In terms of dB.

2) Switch 1 is placed in position 2 to terminate amplifier input terminals. Switch 2 is opened to allow the output noise of amplifier to be further amplified 22 dB by post amplifier. Output noise is read on signal level meter as –20 dBmV.

Actual output noise is calculated by:

Measured output noise = post amp gain + F + input noise + F p + amplifier gain – 20 dBmV = 20 dB + F – 59 dBmV + 22 dB + 7 dB – 20 dBmV = – 59 dBmV + 49 dB + F, F = 10 dB noise contributed.

Now since the noise figure F for a single amplifier has been found, the C/N for a single amplifier can also be calculated.

Example: For an amplifier with a normal carrier input level of +10 dBmV and a noise figure F of 10 dB.

$$Co/No = 59 - F + Cin$$
$$= 59 - 10 + 10 = 59 \text{ dB}$$

Typically CATV amplifiers have noise figures better than 10 dB, usually 4–7 dB.

For the case of identical amplifiers connected together by lengths of cable with a loss equal to each amplifier gain, it was seen that the noise increased by 3 dB every time the number of amplifiers in cascade was double.

$$F \text{ out} = F_1 + 10 \log n$$

F_1 is the noise figure of a single amplifier & N is the number of amplifiers in cascade.

For 2 amplifiers $10 \log 2 = 10(0.301) = 3$ dB.

For 4 amplifiers $10 \log 4 + 10(0.602) = 6$ dB

For 8 amplifiers $10 \log 8 = 10(0.903) = 9$ dB

Using the above mathematical formula the noise figure for a cascade of 7 amplifiers can be calculated. If F_1 for a single amplifier is 5 dB

$$F_{OUT}\Big/_{7 \text{ amplifiers}} = 5 \text{ dB} + 10 \log 7 + 5 \text{ dB} + 8.5 \text{ dB} = 13.5 \text{ dB}.$$

From the example of the cascade of 32 amplifiers, the C/N in dB decreased by 3 dB every time the number of amplifiers in the cascade doubled.

So we can write $\quad C/N\Big/_{\text{cascade}} = C/N\Big/_{\text{single amp}} - 10 \log n$

$10 \log n$ is called the cascade factor.

At the 1st amplifier C/N = 60 dB.

Therefore at the 8th amplifier.

$$C/N_{\text{8th}} = 60 - 10 \log 8 = 60 - 9 = 51 \text{ dB.}$$

This C/N calculation is important because the signal quality of the amplifier cascade depends on a high signal-to-noise ratio.

If the first amplifier becomes noisy, the whole cascade suffers.

Example: For the 32nd amplifier the C/N has degraded to $50 - 15 = 35$ dB.

This C/N is totally unacceptable

A temporary solution would be to switch the first and last amplifier. Now only service on the end will be affected and not much at that.

In a similar proof, the buildup of amplifier distortion can be shown.

Amplifier distortion appears as second- and third-order distortion.

Second-order distortion involves two frequency components such as the second harmonics of a carrier.

Third-order distortion involves three frequency components. There are three types of third-order distortions.

1) Intermodulation $2 f_1 \pm f_2$ $2f_2 \pm f_1$

2) Cross modulation is caused by any false carrier generated by any third-order distortion with another carrier's modulating signal, which affects any of the correct signal carrier. This is similar to another channel's signal affecting the viewed channel.

3) Third harmonic distortion $3f_1$, $3f_2$ etc

Distortion buildup for a cascade.
For second-order distortion, the cascade calculation is like that for noise.

Second-order distortion for the cascade =
second-order for a single amplifier $- 10 \log n$.

For third-order distortion, i.e., cross modulation, single triple beat (3rd harmonic) or composite triple beat.

$$\left. \ast \text{Third-order distortion} \right|_{\text{cascade}} = \text{third-order distortion} \left|_{\substack{\text{single} \\ \text{amplifier}}} - 20 \log n \right.$$

*Either type of third-order distortion.

Manufacturers of cable amplifiers specify all of the single amplifier parameters. Therefore the calculations using the formulas can be used to predict the amplifier cascade performance.

Historically the use of push pull type circuitry all but eliminated second-order distortion. Thus third-order distortion became the limiting factor in setting the maximum number of amplifiers in cascade. Noise buildup is still an important specification.

APPENDIX G

Satellite Ground Station Pointing

The formulas for the earth station azimuth and elevation angles are given as:

$$\text{Az} \quad \text{Angle in degrees} + 180 + \tan^{-1}\frac{\tan D}{(\sin X)} \text{ from true North}$$

$$\text{EL} \quad \text{Angle in degrees} = \left[\frac{[\cos D \times \cos X] - 0.15126}{\sqrt{[\sin D]^2\,[\cos D \times \sin X]^2}}\right]$$

X   = Earth station latitude in degrees
Y   = Earth station longitude in degrees
Z   = Satellite longitude in degrees
D   = Z – Y

Example: Earth station latitude = 41° 38' 11.0" N
           Earth station longitude = 70° 27' 42.0" W
           Satellite longitude = 93.5° W

First convert earth station coordinates to decimal degrees.

Latitude	Longitude
41° = 41.0000	70° = 70.0000
38^1 = 38/60 = 0.6333	27^1 = 27/60 = 0.4500
11.0" = 11/3600 = 0.0031	42" = 42/3600 = 0.0117
41.6364 = 41.63°	70.4617 = 70.46

$$X = 41.64° \quad Y = 70.46° \quad Z = 93.5° \quad D = 93.5° - 70.46° = 23.04$$

For the azimuth angle $= 180 - \tan^{-1}\dfrac{\tan 93.5}{\sin 41.64}$

$$= 180 - \tan^{-1}\frac{0.425}{0.66} = 180° - \tan^{-1}(0.644)$$

$$= 180° + 32.8° = 212.8° \text{ from true North}$$

For the elevation angle $= \mathrm{Tan}^{-1}\left[\dfrac{[\cos 23.04 + \cos 41.66] - 0.15126}{\sqrt{(\sin 23.04)^2 + (\cos 23.04 + \mathrm{sim}\, 46.64)^2}}\right]$

$$= \tan^{-1}\left[\dfrac{(0.920 \times 0.747) - 0.15126}{\sqrt{(0.391)^2\,(0.920 \times 0.664)^2}}\right]$$

$$= \tan^{-1}\left[\dfrac{0.6872 - 0.15126}{\sqrt{0.153 + 0.373}}\right]$$

$$= \tan^{-1}\left(\dfrac{0.5359}{0.723}\right) = \tan^{-1}(0.7389) = 36.5° \text{ from level}$$

Since the azimuth angle is from true North, to use a compass in setting the direction, the magnetic difference (declination) must be known. This value can be obtained from a local surveyor or a topographical map of the area.

Example for a magnetic declination of 15.5°

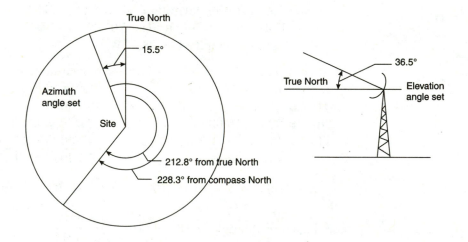

APPENDIX H

Line-of-Sight Microwave Link

The microwave radio link has been used by the telephone industry to transmit and receive telephone traffic for many years. Early systems used the carrier method of single-sideband suppressed-carrier modulation for transmission of a one-way (half-duplex) telephone call. Another carrier operating on another carrier going the other direction carried the other half of the call. This technique is known as *frequency division multiplexing* (FDM). Later, telephone voice signals were converted to HRZ digital pulse streams, which is referred to as *time division multiplexing* (TDM). Both methods were used to stack voice channels on one RF microwave carrier. It is this carrier that is of major concern and should be received at the receiving site in the best possible condition. The basic microwave link for one path is shown below.

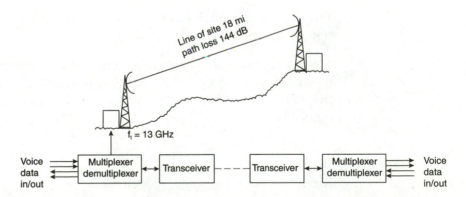

Path length L = 18 miles path loss = 36.6 = 20 log L + 20 log f_t dB

Path loss 36.6 + 20 log 18 + 20 log 13000 = 36.6 + 25.1 + 82.3 = 143.98 = 144 dB

This loss assumes nothing is in the path way and there is no signal refraction.

To examine the effects of refractions, the causes will be investigated. For a *line-of-sight* (LOS) microwave radio link, objects interfering with the line-of-sight path will refract or diffract the transmitted beam. Since the earth is spherical, the transmitter antenna and receiving towers have to be of sufficient height to clear the earth's bulge. The figure below illustrates this concept.

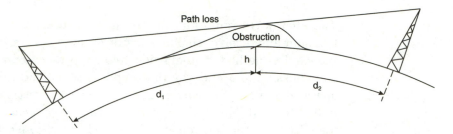

The earth's bulge in feet is given by:

$$h = 0.667 \, d_1 \, d_2 \qquad \begin{array}{l} d_1, d_2 \text{ in miles} \\ h \text{ in feet} \end{array}$$

$$\text{or } h = 0.079 \quad d_1 \quad d_2 \qquad \begin{array}{l} d_1, d_2 \text{ in km} \\ h \text{ in meters} \end{array}$$

The bulge of the earth has the effect of raising the height of the obstruction. If the earth had no bulge, then of course $h = 0$.

The earth is not purely spherical, and the atmosphere gets thinner with height. This has the effect of changing the earth's radius. The X factor is the ratio of the effective earth's radius to the true earth radius. Mathematically

$$K = \frac{\text{Effective earth radius}}{\text{True earth radius}}$$

K can be negative, infinite, or positive.

When K is – the beam path is concave upward
$K = \infty$ the beam path follows the earth's curvature
For small values of K<1, beam is bent upward
$K = 1$ beam follows optical line of sight.
Temperature also affects K. K is related to h by the expression:

$$h = \frac{0.667 \, d_1 \, d_2}{k}$$

A nominal value of k used is $k = \dfrac{4}{3} \ 1.33$

Fresnel zones

The wave front from the transmitting antenna expands as it travels through space, which results in reflections and phase changes as the wave passes over obstacles & obstructions. The height of the transmitting & receiving sites should allow for extra height. The beam cross section shows first, second, and third fresnel zones as concentric bands around the beam center axis, as shown below:

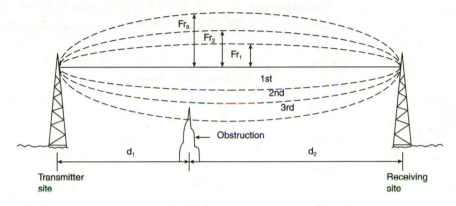

$$Frn = 72.1 \sqrt{\frac{nd_1\,d_2}{fD}}$$

d_1 = Distance in miles from transmitter to obstruction

d_2 = Distance in miles from receiver to obstruction

f = Operating frequency in GHz

n = Fresnel zone number

Frn = Fresnel zone radius in feet

D = $d_1 + d_2$ in miles

For metric units d_1, d_2 And D in km

                                      f in GHz and Frn in meters

The formula becomes

$$Frn = 17.3 \sqrt{\frac{nd_1\,d_2}{f\,D}}$$

These formulas are used to adjust tower heights and site elevations so objects or obstructions do not penetrate the 1st fresnel zone.

Path profiling

The procedure of path profiling is to make sure the transmitting and receiving site are clear of obstructions and the path or beam, taking into account the fresnel zones, is far enough above the earth's surface to be unaffected. This procedure might require some on-site work with surveying instruments. However, reasonable estimates can be made using 7.5 min topograhical maps and appropriate graph paper. Also sea-level refractivity contour charts as well as K factor versus refractivity might be helpful in estimating the K factor for the area. A short example illustrates this procedure.

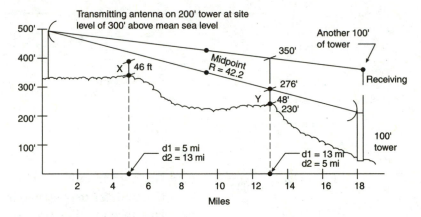

L = 18 mile hop

For the obvious obstruction points X & Y with K = 0.95

Earth's curvature (bulge) at point X $hx = \dfrac{0.667 \times 5 \times 13}{0.95} = 45.6 = 46$

$hy = \dfrac{0.667 \times 13 \times 5}{0.95} = 46$ (Note we are right at beam center at point Y.)

Now the fresnel zone radius must be calculated. This is done for the mid point using the following formula:

$$R = 72.1 \sqrt{\dfrac{25}{13 \times 18}}, \quad R = 72.1 \sqrt{\dfrac{d_1 d_2}{f_{GHz} Lm_1}} = 72.1 \sqrt{\dfrac{81}{13 \times 18}} = 42.4 \text{ ft}$$

To correct this figure for points X and Y, we can use the formula or the graph in the following figure.

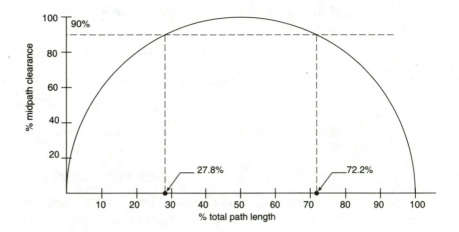

At point X d_1 = 5 mi   Which is $\dfrac{5}{18} \times 100\% = 27.8\%$

At point Y d_1 = 13 mi   Which is $\dfrac{13}{18} \times 100\% = 72.2\%$

So at point Y fresnel radius is 0.90 of mid span value of 42.4 ft or 0.90×42.4 ft = 38 ft. At this point the bulge of the earth penetrates the fresnel zone and no clearance results.

Another 100 ft of receiving tower height should do the job.
At Y elev. with h = 46' + 230' = 276'
  Corrected fresnel radius = 38

276' + 38 = 314' + path line of site elevation at point Y = 350' so there is clearance allowances for vegetation. Trees might make an additional 50 ft of receiving tower height more prudent.

Once the microwave path is cleared, recall that the 18-mile path loss was 144 dB = PL.

Suppose this was a high-power transmitter with an output power of 1 W converted to dBm. Recall that 1 mW corresponds to 0dBm.

$$P_T = 10 \log \frac{1w}{1 \times 10^{-3w}} = 30 \text{ dBm}$$
$$\phantom{P_T = 10 \log \frac{1w}{1 \times 10^{-3w}} = } 1\text{mW}$$

If the gain of the transmitting antenna is 45 dB, then the radiated power P_R is calculated.

$$P_R = 30 \text{ dBm} + 45 \text{ dB} = 75 \text{ dBm}$$

At the receiving antenna we may calculate

$$P_{REC} = P_R - P_L = 75 - 144 = -69 \text{ dBm}.$$

If the gain of the receiving antenna is 40 dB, then the receiving power P_{REC} is calculated as

$$P_{REC} = -69 + 40 = -29 \text{ dBm}.$$

If the receiver manufacturer specifies the noise power level as –90 dBm, then the receiver input carrier-to-noise ratio is calculated as:

$$C_R/N_r\Big/_{dB} = C_{R_{dB}} - N_{R_{dB}} = -29 - (-90) = 59 \text{ dB} = 60 \text{ dB}$$

The carrier level is 60 dB greater than the noise level.

A good receiver should be able to produce a baseband S/N of about 52 dB with a 60-dB carrier-to-noise ratio specification.

The fade probability of an outage occuring based on Rayleigh fading the worst month of the year can be calculated by:

$$P_F = T_F \times T_{TF} \times 2.5 \times 10^{-6} \times f \times L^3 \times 10^{-\frac{fm}{10}}$$

P_F = Fade probability of an outage occurring
T_F = Terrain factor   4 for smooth terrain & water
                                             1 for average terrain.
                                               0.25 rocky mounting terrain
T_{TF} = Temperature/humidity factor   0.5 for hot/humid areas
    f = Operating frequency in GHz   0.25 normal climate
    L = Length of path in miles   0.125 high mountain & dry areas
   f_m = System fade margin 35 dB AML
                                    30 dB others

Example: Normal terrain/normal climate for the 18-mile hop at 13 GHz and 30-dB fade margin

$$P_f = 1 \times 0.25 \times 2.5 \times 10^{-6} \times 13 \times 18^3 \times 10^{-\frac{30}{10}}$$
$$= 8.1 \times 10^{-6} \times 18^3 \times 10^{-3}$$
$$= 8.1 \times 10^{-6} \times 5832 \times 0.001$$
$$= 47239 \times 10^{-6} \times 0.001$$
$$= 0.047239 \times 10^{-3}$$

For a month consisting of 720 hrs

$$\text{Amounts to } 0.0472 \times 10^{-3} \times 720 \text{ hrs} = 30.5 \times 10^{-3} \text{ hrs}$$
$$= 0.031 \text{ hrs}$$
$$= 0.031 \text{ hrs} \times 60 \text{ min/hr}$$
$$= 1.86 \text{ minutes in a month}$$

Heavy rain can affect the line-of-sight microwave link. At frequencies below 10 GHz, rain does not have much effect. Studies of rainfall over a five-year period can indicate seasonal rains that could cause problems.

Bibliography

Bartlett, Eugene R. *Cable Television Technology and Operations*. ISB7-0-07-003957-7. New York: McGraw-Hill, Inc., 1990.

Bartlett, Eugene R. *Cable Communications*. ISBN 0-07-005355-3(H). New York: McGraw-Hill, Inc., 1995.

Benson, K. Blair, ed. *Television Engineering Handbook*. ISBN 0-07-004779-0. New York: McGraw-Hill, Inc., 1996.

Chomycz, Bob. *Fiber Optic Installations*. ISBN 0-07-011635-0. New York: McGraw-Hill, Inc., 1996.

Digi Points. Vol. I, Issue 8. August, 1997. SCTE Interval.

——————. Vol. II, Issue 6/98. SCTE Interval.

Grant, William O. *Cable Television*, 2nd ed. IXU-295-120. Fairfax, VA: GWG Associates, 1998.

Hecht, Jeff. *Understanding Fiber Optics*, 3rd Ed. ISBN 0-13-956145-5. Upper Saddle River, NJ: Prentice Hall, 1999.

Hioki, Warren. *Telecommunications*, 3rd Ed. ISBN 0-13-632043-0. Upper Saddle River, NJ: Prentice Hall, 1998.

Hunting, Shannon. "Using TCM Part II." *R. Howald Communication Systems Design Magazine*. April, 1999.

Inglis, Andrew F. *Video Engineering*. ISBN 0-07-031716-X. New York: McGraw-Hill, Inc., 1993.

Miller, Gary M. *Modern Electronic Communication*. ISBN 0-13-217879-6. Englewood Cliffs, NJ: Prentice Hall, 1996.

Noah, Jeffery. *FCC Required Measurements for U.S. Cable Systems*. (Application Note.) Beaverton, OR: Tektronix, Inc.

Ramteke, Timothy. *Networks*. ISBN 0-13-958959-X. Upper Saddle River, NJ: Prentice Hall, 1994.

Raskin, Donald and Dean Stoneback. *Broadband Return Systems for Hybrid Fiber / Coax Cable TV Networks*. ISBN 0-13-636515-9. Upper Saddle River, NJ: Prentice Hall.

Robin, Michael and Michel Poulin. *Digital Television Fundamentals*. ISBN 0-07-053168-4. New York: McGraw-Hill, Inc., 1998.

Simons, Ken. *Technical Handbook for CATV Systems*, 3rd Ed. Hatboro, PA: Jerrold Electronics Corporation (General Instrument), 1978.

Sterling, Donald J., Jr. *Technician's Guide to Fiber Optics*, 2nd Ed. ISBN 0-8273-5835-0. Albany, NY: Delmar Publishers, 1993.

Thomas, Jeffrey L. *Cable Television Proof-of-Performance*. ISBN 0-13-573726-5. Englewood Cliffs, NJ: Prentice Hall (IEEE Press), 1995.

Tomasi, Wayne. *Electronic Communication Systems*. ISBN 0-13-220021-X. Englewood Cliffs, NJ: Prentice Hall, 1994.

Vergers, Charles A. *Handbook of Electrical Noise Measurement & Technology*, 2nd Ed. ISBN 0-8306-2802-9. Blue Ridge Summit, PA: TAB Books, 1987.

Winch, Robert G. *Telecommunication Transmission Systems*. ISBN 0-07-070964-5. New York: McGraw-Hill, Inc., 1993.

Glossary of Current Commonly Used Acronyms

Term	Meaning
8-VSB	8-Vestigial Side-Band
ACATS	Advisory Committee on Advanced Television Service
ADSL	Asymmetrical Digital Subscriber Line
AMI	Alternate Mark Inversion (Digital Signal Coding)
ASCII	American Standard Code for Information Interchange
ATM	Asynchronous Transfer Mode
B8ZS	Binary Eight-Zero Suppression
BPSK	Binary Phase Shift Keying
BOC	Bell Operating Company
CAP	Competitive Access Provider
CCIR	International Radio Consultative Committee
CCIS	Common Channel Interoffice Signaling
CCITT	International Telephone and Telegraph Consultative Committee
CO	Central Office
CPU	Central Processor Unit
CRC	Cyclical Redundancy Checking
CSMA/CD	Carrier Sense Multiple Access Collision Detection
DACS	Digital Access and Cross-Connect System
DCE	Data Communications Equipment
DDD	Direct Distance Dialing
DDN	Digital Data Network
DTE	Data Terminal Equipment
DLC	Digital Line Carrier
DOCSIS	Data Over Cable Service Interface Specification
DOS	Disk Operating System
DOV	Data Over Voice
DS-0	Digital Signal, Level 0
DS-1	Digital Signal, Level 1

DSS	Direct Satellite Service
DSP	Digital Signal Processing
DSX-1	Digital System Cross-Connect
DOCSIS	Data Over Cable Service Interface Specification
DVD	Digital Video Disc
DVOD	Digital Video On Demand
FCC	Federal Communications Commission
FDDI	Fiber-Distributed Data Interface
FDM	Frequency Division Multiplex
FDMA	Frequency Division Multiple Access
FSK	Frequency Shift Keying
GPS	Global Positioning System
HDLC	High-Level Data Link Control
HFC	Hybrid Fiber Coax
IEEE	Institute of Electrical and Electronics Engineers
IP	Internet Provider
ISDN	Integrated Services Digital Network
ITU	International Telecommunications Union
JPEG	Joint Photographic Experts Group
LAN	Local Area Network
LED	Light-Emitting Diode
MPEG	Motion Picture Experts Group
MSO	Multiple System Operator
OSI	Open Systems Interconnection
PBX	Private Branch Exchange
PCM	Pulse Code Modulation
PCS	Personal Communications System
POD	Point of Deployment
POP	Point of Presence
POTS	Plain Old Telephone Service
PSTN	Public Switch Telephone Network
QAM	Quadrature Amplitude Modulation
QPSK	Quaternary Phase Shift Keying

SCTE	Society of Cable Telecommunications Engineers
SDH	Synchronous Optical Network
SDLC	Synchronous Data Link Control
SMPTE	Society of Motion Picture and Television Engineers
SONET	Synchronous Optical Network
SPE	Synchronous Payload Envelope
TDMA	Time Division Multiple Access
VPI	Virtual Path Identifier
WAN	Wide Area Network

INDEX